THE B-24 LIBERATOR

SD
4

ALLAN G. BLUE

THE B-24 LIBERATOR

A Pictorial History

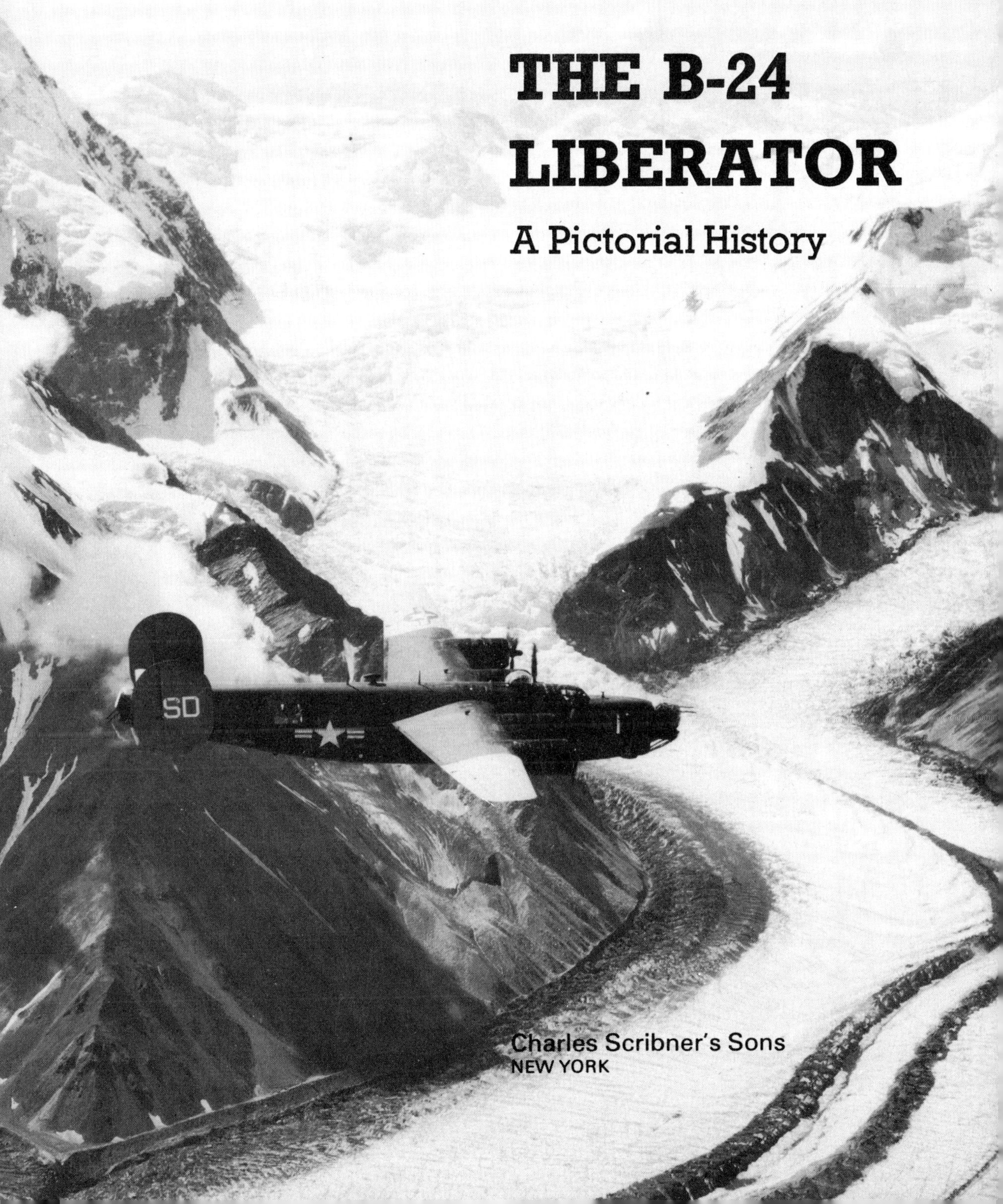

Charles Scribner's Sons
NEW YORK

This book published simultaneously in the United States of America and in Canada.

Printed in Great Britain
Library of Congress Catalog Card Number 75-21664
ISBN 0-684-14508-1

Title page: PB4Y-1's of VP-61 in Alaska, 1948. / USN

Below: Kingman, Arizona, February 1947. / W. T. Larkins

Contents

Acknowledgements

The nature of this study of the Liberator is such that its primary contributors cannot be named here. They include the unknown writers and illustrators who produced the voluminous technical manuals on the Liberator and its many offspring, as well as the equally anonymous authors and compilers of the procurement documents, the parts lists, the USAAF Technical Orders, the General Service Notes, the Flight Test Reports, the operational theatre evaluations . . .

And then there are those unknown heroes—or were they heroines?—who so tediously but carefully hand-posted the aircraft record cards maintained at Wright Field for each individual B-24 the USAAF ever accepted.

To all of these nameless individuals, as well as to those who have seen to it that the results of their labours have been preserved for three decades, must go the credit for making this history possible.

The long road that stretched between possibility and reality, however, was traversed with the help of individuals who *are* known and who should be known to the reader as well. Without their assistance and encouragement the journey would have been much more difficult indeed. They include:

Hal Andrews, Roger Besecker, Steve Birdsall, Bob Brooks, Lew Casey, Bill Chana, Jim Fahey, Joe Famme, Royal Frey, George Gillberg, Gordon Jackson, Frank Green, I. M. Laddon, Ed Lazarras, Jack Love, Bob Lutz, Bernie Mallon, Dean McCoy, Bob McGuire, Lew Nalls, Lee Pearson, Bill Plant, John Preston, Phil Proffitt, Kenn Rust, Tony Shennan, Ken Sumney, Bob Webb, John Wible and Bob Wood.

Special thanks are due the Convair Division of General Dynamics Corporation for their generous co-operation and assistance.

Without the help of Roger Freeman the manuscript would, in all likelihood, still be in that form rather than in this.

Finally, I acknowledge the patient assistance of Mrs Wesley Graham, who by this time probably knows more about the Liberator than any other woman alive. A dubious distinction for her, perhaps, but an invaluable one to me.

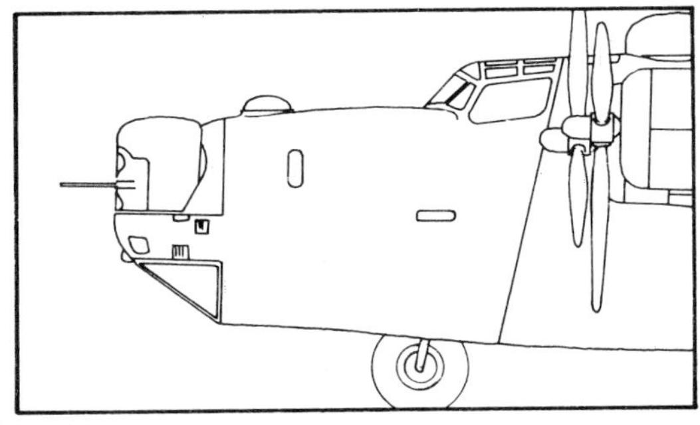

Introduction

Not many more than a handful of people had seen or heard of the Consolidated Liberator before December 1941, and by December 1945—only four years later—it had virtually passed, unmourned, into history. Yet in that brief span of time it had been produced by the thousands, had borne tens of thousands into the skies and had been seen there by millions of people in every part of the world. Depending upon when you listened, during those epic four years, you could have heard that the Liberator was the white hope of the Allied Air Forces, a killer ship that pilots were scared to death of, a superb and versatile weapon unmatched in America's air arsenal, or a hard-to-handle, underpowered and over-rated flying bomb. In reality and at times it was all of these, and more.

In many ways it was a true mercenary, taken on in large numbers to fight a dirty, brutal but necessary war in which it served with sometimes spectacular, sometimes disastrous—most often just ruthlessly repetitive—effectiveness. Liberator units took their losses and inflicted even greater losses in return, overwhelming the enemy with numbers when no other tactic would prevail.

It was equally in its element as a solitary sentinel ranging far at sea to search for a German U-Boat or to photograph a Japanese-held atoll, or as a member of the multitudinous bomber formations that streamed across the Channel and the Alps to carry the war to the German heartland. In other versions it tied the Hitler War and the Hirohito War together by transporting high-priority men and material wherever they were needed most at the moment.

When evaluating any personal opinion of a particular aircraft one cannot over-stress the importance of individual attitude. Take 1st Lt Earl S. Kimball, for example. En-route to the European Theatre of Operations (ETO) as replacements, Lt Kimball and his crew were trying to locate and land at Marrakech, French Morocco, in darkness when their B-24 brushed the side of a mountain. The impact tore 7 ft off one wing, including several feet of aileron, but Kimball managed to keep the plane in the air until he could find a level spot to put it down. This he did without injury to the crew or further damage to the Liberator. When the cold light of day revealed that he had landed in such desolation that a road would have to be built to the plane in order to repair it, Kimball dismissed all of his crew except the co-pilot and proceeded to fly the disabled ship out of the emergency landing spot and on to Marrakech. This was considered quite a feat in itself, but in subsequent weeks Kimball made *five additional* flights in his 'wingless wonder'. These were to various points in Africa in search of a new outer wing panel so he and his crew could get on with their original mission.

Nobody can tell Earl Kimball that the B-24 wasn't a damn good aeroplane.

Or take Group Captain John A. Powell, RAF. On 1 October 1943, the B-24 in which Group Captain Powell was flying as observer was severely damaged by anti-aircraft fire over Wiener Neustadt. A direct hit in the nose of the aircraft killed the bombardier and seriously wounded the navigator. When a succeeding burst shattered the leg of the pilot, wounded the co-pilot, and started a fire in the bomb bay, Group Captain Powell took over as pilot. Although he had never flown a B-24, two engines were out and the controls badly damaged, Powell brought the crippled Liberator back. Then, at an altitude of 250 ft and beyond gliding distance from the nearest landing ground, the fuel supply failed and the two remaining engines had to be switched off. Group Captain Powell then crash-landed successfully, in darkness, without further injury to the B-24's occupants.

That crew thought pretty highly of the Lib, too.

In spite of the foregoing, this book is not a eulogy. Far from it. There are quite a few unkind words about the old bird in the pages that follow—nearly all of them richly deserved. Some men did hate the Liberator, or feared it, or both. A few loved it. Between these extremes there was the majority—that neither hated nor loved the aircraft but flew in it and fought in it because it was the means that had been given them to achieve an end. These men learned to live and fly with the ship's shortcomings and to make the most of its attributes. Their record proved beyond a doubt that the Liberator could do, and did well, the job for which it was created.

The purpose of this book is to provide, within the space available, information that will answer most of the questions that from time to time arise among aerophiles concerning the design, development, production and operational use of the Liberator/Privateer family of aircraft. To further this objective a large amount of data is given in tabular form because that is the most efficient way to present it and also the most convenient reference format. With few exceptions this data is not duplicated in the text. The author wishes it were possible to say that all such statistics have been relegated to the back of the book, but such is far from the case. For as the biography of a personage must include some scale against which to measure the progression of its subject through time, so must the story of an aircraft at times be filled with the serial numbers, block numbers and production numbers which mark the growth and development of its subject.

The reader well versed in Liberator lore will note that many statements of fact presented herein are at variance with data contained in previously published articles and monographs on this aircraft. Suffice it to say that the author, also aware of these discrepancies, has made doubly sure of his documentation in these instances and, in fact, in many cases has been able to locate the probable source of the erroneous information that appeared in early Liberator literature, both official and unofficial, and has been perpetuated ever since. While repetition gives a certain acceptability to spurious information it cannot, for example, change the fact that there was only one YB-24 built, not seven. Or that the B-24H, not the B-24G, was the first Liberator with a production nose turret. Or that Consolidated, not Ford, produced the XB-24K. This is in no way meant to belittle the efforts of those earlier chroniclers; without them we would know much less about the Liberator than we do. They have told the basic story, to which we can only add refinements.

As for the errors in this volume—the existence of which must be rated as certain as oil stains around a B-17's nacelles—they in turn will be ferreted out and the correct facts brought to light. In advance apology, the author can only plead guilt without premeditation. With the help of a great number of wonderful people whose assistance is acknowledged elsewhere within these covers, we've done our best to tell it as it was.

Right: Beauty and the Beast. /General Dynamics

RUGGED BUT RIGHT
GALS

8 October 1942

'Briefly, the situation is this: The B-17 is a fine, heavy bomber which has been lavishly built up by the Press with the result, we believe, that not only the public but the personnel in the Army Air Forces think of it as an airplane far superior to any other heavy bomber. At the same time, our industry is just beginning to put out large numbers of B-24's. Even in its condition today, without the (lower) turret, which may be available in quantity by the first of the year, the B-24 has shown up in proving ground tests as a very fine heavy bomber with a greater range than the B-17.

'Brereton has used this same B-24 with German and Italian opposition and has had a remarkable degree of success in air combat. Likewise, Butler's small B-24 unit has been highly successful against the Japs in the Aleutians. It is unfortunate at this time that neither of those theaters has had the publicity enjoyed by the B-17 in the United Kingdom. The net result is a false public impression that the B-17 is a fighting airplane far superior to any other heavy bomber in the world, because of the briefness of B-24 combat experiences and lack of publicity for its successes in battle.

'We find ourselves faced with what may be a real and acute problem in psychology and in leadership.'

H. H. ARNOLD,
Lieutenant General, USA
Commanding General, Army Air Forces

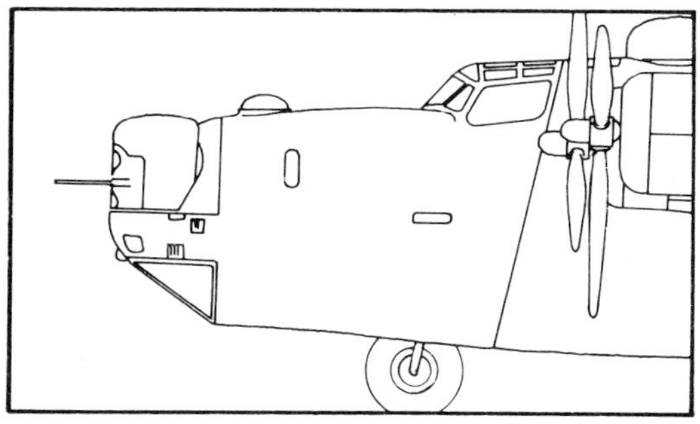

I
Design, Production and Allocation

Early History

In the late 1930s the Consolidated Aircraft Corporation (CAC) of San Diego, California, was one of the leading companies producing long-range flying boats for the US Navy and, as such, was a logical firm for a man named David R. Davis to approach. Davis was no stranger to aviation, having put up the money for Donald Douglas's first aeroplane in 1920. Later he worked on a variable pitch propeller for Bendix, until the stock market crash in 1929 ended the Bendix work and badly mangled Davis's personal finances. Now, in the summer of 1937, his main asset was US Patent No. 1942688, filed 25 May 1931 and issued 9 January 1934, which related to the development of aerofoils by the use of certain mathematical formulae. Using this method he had arrived at a family of wing sections which he claimed gave appreciably better performance than any aerofoil then in use, and which he felt would be particularly applicable to the long-range flying boat field. Already turned down by one major US company, Davis had obtained an interview with the president of Consolidated, Major Reuben H. Fleet, to present his ideas.

Fleet wasn't impressed, nor was his chief designer, I. M. Laddon, who was called in to listen. The scepticism was understandable, for the National Advisory Committee for Aeronautics (NACA) had, over an extended period, conducted a most comprehensive and systematic study of aerofoil shapes, the results of which did not include anything similar to Mr Davis's wing. However, a few days later Laddon told Fleet he had a hunch that they should give the Davis formula wing a try. Still not convinced, Fleet dubiously agreed to pay for construction of a model and a wind tunnel test at the California Institute of Technology. Consolidated had a new flying boat on the drawing boards. They would let Davis have a try at designing a wing for it and then test his offering against their own wing design—which Consolidated felt in itself was near the ultimate for aircraft of the type under consideration.

The results of the wind tunnel tests were unbelievable—so much so that the Cal Tech engineers re-calibrated their tunnel and ran the tests a second time, and then a third time, to make sure. Finally they delivered a report to Fleet which said, in part, that the 'remarkably high value for the Davis wing is probably associated with a peculiar variation of boundary layer thickness with angle of attack, but no real explanation for it has yet appeared'. In over-all tone, the report left the impression that the wing could be a wind tunnel fluke, or it just might be a really revolutionary concept.

After what must have been a considerable amount of soul-searching, Fleet chose to gamble. The Davis wing would go on the twin-engined Model 31 flying boat, a CAC private venture which was planned both as a commercial transport and a possible successor to the PBY Catalina. And, if it worked out successfully, it might also be used on a projected land-based bomber for which a number of design studies were in progress.

On 9 February 1938 Davis entered into a licence agreement with Consolidated whereby he would receive $2,500 for each prototype and a royalty on each subsequent aircraft produced using his aerofoil, based on a sliding scale beginning at ½ of 1% of the selling price of the aircraft (less engines, propellers, other items of government furnished equipment, and spares) and decreasing to 1/8 of 1% when and if orders reached $10 million. In addition, when total royalty payments reached $50,000, the rate was to be reduced to 1/16 of 1%.

During the latter part of 1938 the US Army Air Corps, not unaware of the growing possibility of war in Europe, asked Consolidated to consider becoming a second source for B-17 production. San Diego personnel dutifully visited the Boeing plant in Seattle to study the proposal, but their report to the Army rejected the idea on the basis that the B-17 design was incomplete and, in any case, would be hard to adapt to CAC building methods. What CAC had in mind, of course, was something far more ambitious than becoming a second factory turning out copies of what they had come to refer to as Boeing's 'Hollywood Bomber'. In early January 1939, Mr Frank W. Fink was called to Major Fleet's office, where Fleet and Laddon told him they had decided that Consolidated Aircraft could build a better bomber than the B-17 and in as short a time as it would take to become a second source for that aircraft. In Fink's words, 'They told me that I was to be project engineer and they wanted a mock-up of the new bomber to be built in two weeks. They were planning on leaving for Wright Field to convince the Army to build this bomber and to have them appoint a mock-up board immediately to come out and see the mock-up. My first question was, "What does this bomber look like? Which one of the proposals that we've worked on during last year?" They told me that there was no drawing but they would describe it to me. The description given was to use the Davis wing from the Model 31, a twin tail from the Model 31, four engine nacelles from the PBY Catalina, and build a new fuselage which had two bomb bays, each the same size as the B-17's one bomb bay. I was therefore given the job of getting the required number of sketches out of a "verbal" design, and a mock-up built by the Experimental Shop, in fourteen days!'

Two weeks later, Fink received a phone call from Major Fleet, who was at Wright Field. The mock-up board was assigned; was the mock-up ready? Fink told him it was completed and being painted at that moment. The mock-up board came out within a few days, with Lt Donald L. Putt in charge. While the board stayed in San Diego, many changes were made in the fuselage, principally in gun placements, windshield arrangements, and the bombardier's station. Before the board left, CAC had built a second complete fuselage mock-up, which then became the design used for the XB-24.

In order to satisfy procurement regulations, a US Army Air Corps (AAC) Type Specification (No. C-212) was drawn up and, on 1 February 1939, the Glenn L. Martin Company and Sikorsky Aviation Corporation were solicited for possible designs which would meet the Specification requirements. Since less than three weeks were allowed for reply, the gesture was a mere formality and on 21 February 1939 the Consolidated proposal was recommended to Washington for approval. The resulting Army contract was formally signed on 30 March 1939. It ratified the preliminary work already done, calling for one wooden mock-up, one wind tunnel test model, and one XB-24 aeroplane—with the latter to be completed by 30 December of the same year. The proposed aeroplane officially became the CAC Model 32 at this time.

The Model 31, upon which work had begun first, made its inaugural flight on 5 May 1939, and the enthusiastic report of Consolidated's chief test pilot, William Wheatley, left no doubt that Fleet's gamble on the Davis wing had paid off. Now the Consolidated designers proceeded with no misgivings with regard to control or aerodynamics of the Model 32, since the wing and empennage design for the XB-24 were borrowed directly from its sister aircraft. In fact, in working his fourteen-day miracle Fink utilised almost all of the wing and tail portions of the actual mock-up previously prepared for the Model 31.[1]

The XB-24, assigned Army Air Corps serial 39-556, first flew on the afternoon of 29

Above: The Model 31. The story goes that CAC, looking ahead to a possible Army contract, refused to incorporate Navy-desired Catalina type wingtip floats in order to preserve the aerodynamic integrity of the Davis wing for future application to a land-based bomber. / General Dynamics

Left: I. M. Laddon. / General Dynamics

Below: XB-24 utilized PB4Y-5 type nose cowls. Intake at top was for carburettor air. / Pratt & Whitney

Top: The first Liberator. As XB-24 (39-556) in flight on 27 February 1940, with slots (five on each wing) just barely visible outboard of the national insignia and pitot tube on each wing.

Above: As XB-24B (39-680) with new engines, which appeared for a short time on the spinners shown here. New serial has been stencilled on nose.

December 1939, one day under the contract deadline. The inaugural flight lasted seventeen minutes, with CAC test pilot Wheatley at the controls. The first AAC personnel to pilot the aircraft were Major Umstead and Captain Harmon, who took it aloft on 17 February 1940, although the first official Army flight test did not take place until 18 March. On one of the very early flights the up-latch mechanism failed to secure one leg of the main landing gear. The hydraulic system kept the gear retracted, but with enough movement that the flight crew thought wing flutter might be present. The problem was quickly located after landing and the tests continued.

By this time additional US orders had been placed, so that production planning was already well under way for these and for an export version (the LB-30) that the French, and later the British, had added to the backlog.[2] Then as now, however, it took time to tool up for series production and it would be December of 1940 before the first production article was accepted from the San Diego plant.

The origin of the designation LB-30 is rather complicated. The French order—Contract A.F. 7—was for 175 aircraft to be designated LB-30MF's. In 1961 Mr Laddon recalled, in a letter to the author, that the LB stood for *Land Bomber* and the MF stood for *'Mission Francais'* or French Mission, which ordered the aircraft. This, however, did not explain the '30', since the Liberator was obviously in no way related to the Consolidated Model 30 or XPB3Y-1 which, although never built, was to be a large patrol bomber for the USN with a 169 ft span, a length of 104 ft and powered by four R-2800-18 engines. The answer continued to elude this writer for eleven years, but is now known to lie in the fact that Consolidated kept two sets of books—one for design proposals and another for models, with the latter assigned only to those designs intended for actual production. Design proposals were numbered sequentially, with each given a two-letter prefix to designate the aircraft type—e.g., LB for Land Bomber, LC for Land Commercial, SC for Seaplane Commercial, etc. Model numbers were also assigned sequentially, but bore no prefix and no relationship to the design proposal numbers from which they stemmed.

Between 26 August 1938 and 7 March 1940 CAC produced twenty 'LB' design proposals, starting with the LB-4 and ending with the LB-29[3]. At that point the separate design proposal identification system was dropped, and no design proposal numbered beyond LB-29 appears in CAC's records. The French contract negotiations, however, were nearing completion during March-April 1940 and the sale required the assignment of a Consolidated design proposal number to the 'export' version of what had become, by this time, the Model 32 for the US Army. CAC simply chose the next number that would have been used under the old system—LB-30.

While plans for production went forward development continued on the XB-24, which had failed to meet the top speed specified by the Army under the original contract. Wing slots were found to be unnecessary and were deleted. The pitot tubes were relocated from the wing to the forward fuselage, and 2 ft was added to the tail span. Additional wind tunnel testing examined the possible benefits of wing and tail fillets, re-design of the engine nacelles and a fully retractable main gear instead of the existing partially exposed and faired configuration. In the end, time was the controlling factor and only those changes which could be made quickly were permitted.

Designer Laddon had chosen to use the Pratt & Whitney R-1830 engine over the contemporary Wright R-1820 because of the smaller frontal area—and therefore reduced drag penalty—of the twin-row P&W design. The mechanically-supercharged (R-1830-33) version installed in the XB-24, however, did not have the high altitude performance of the forthcoming turbo-supercharged R-1830-41, which would soon be available in quantity and which the Army perferred for its machines along with the addition of self-sealing gasoline tanks and protective armour plate for the crew.

All of these changes were incorporated into the XB-24, which then became the XB-24B. This aircraft was assigned a new serial number, 39-680, and finally accepted by the Army on 13 August 1940. (Serial number 39-556

Left: Main landing gear strut for the XB-24. Although the outward-retracting Liberator main gear had a somewhat fragile appearance, the design was sound and, except for some strengthening added during block D-15-CO, remained practically unchanged throughout production.

Below: Bottom view of original 1/36-scale XB-24 wind tunnel model as modified in June 1940 to test means of increasing Liberator top speed. Nacelles have been redesigned, fillets added to wing and empennage, and wheels and wheel fairings removed to simulate full main gear retraction. The original smaller-span horizontal stabiliser of the XB-24 is seen on this photograph.

Right: XB-24B flight test. Note test fairings installed at junction of wing and number two engine nacelle. / General Dynamics

Below right: Photo taken the day after Christmas, 1940, shows camouflaged LB-30A's in background while work proceeds on LIB I's. / General Dynamics

was cancelled.) Although the fuel cells which were fitted inside the original XB-24 integral tanks reduced the fuel capacity by approximately 300 gallons in each wing, the Air Corps wanted the new tanks and, even more so, the newer engines.[4] As a result, six YB-24's and twenty B-24A's were 'borrowed' from Army contracts and delivered against the British order so that the US could wait for the later, improved machines.

First Production

Because of the contract exchange, the first of the long line of production Liberators to be rolled out were the six LB-30A's (ex AAF YB-24's) for the British, all of which were accepted in December 1940. The first LB-30A was test-flown on 17 January 1941. Lacking self-sealing tanks but possessing great range, these aircraft, bearing British serial numbers AM258/263, were employed on the Trans-atlantic Return Ferry Service between England and Canada after suitable modifications had been carried out at Montreal. Besides the removal of armament, these changes ultimately included cabin heat ducted from the engine exhaust system, crew and passenger oxygen, oil cooler relocation, and provisions for carburettor heating, alcohol engine and propeller de-icing, and a fuel dumping capability.

Next came the twenty Liberator I's (AM910/929), beginning with two in March, six in April, eleven in May and one in June 1941. Carried as LB-30B's or 'B-24A Conversions' on Consolidated records, one was test flown by Consolidated with large, rather blunt spinners on the two port engines while the starboard engines retained the normal hub arrangement for comparison purposes. The Liberator I carried twin ·30 calibre machine guns in the tail and single ·30 calibre guns in side and belly apertures. However, this armament was augmented in various combinations by the British (e.g., six 20 mm cannon and four depth charges) before most of the Liberator I's were put into service with Coastal Command in the summer of 1941. Four were allotted to BOAC.

May 1941 also saw acceptance of the first and, as it turned out, only YB-24 (40-702). Identical to the Liberator I except for armament, which substituted ·50 calibre guns in all but the tail position, this aircraft was received on 22 May with camouflage finish. Immediately following came eight B-24A's in June (40-2369/2376) and one more (40-2377) in July.

Also indistinguishable in outward appearance from the Liberator I, most of these aircraft were assigned transport duties by the USAAF. Two of them, 40-2373 and 40-2374, flew members of the Harriman Mission to Moscow in September 1941.

Two more were specially equipped and scheduled to be sent on a secret mission to secure photographs of the Japanese 'mystery' atolls of Jaluit, Truk, and possibly Ponape, while ostensibly on their way to the Philippines. One of these aircraft (40-2371, Lt Ted S. Faulkner) arrived at Hickam Field, Oahu, on 5 December 1941 and was awaiting additional armament, to be brought from the US by the second aircraft, 40-2375, when the Japanese attack on 7 December rendered execution of the plan impossible and destroyed Lt Faulkner's B-24 where it was parked inside one of Hickam's hangars. The other Liberator was held in the States and scheduled for 'Project X' (see below) but was withdrawn at the last moment and turned over to the Ferrying Command with which it served until after the war. In fact, it was exactly four years after the Pearl Harbor attack that this aircraft, which might have itself opened hostilities against Japan in World War II (the photo mission crews were to be instructed, if attacked, to 'use all means in your power for self preservation'), was ignominiously reduced to scrap at Walnut Hill, Arkansas, after logging over 10,000 hours in ATC service where it was known, appropriately, as '*Old Consistent*'.

In August 1941 the first Liberator II was accepted at San Diego. One hundred and thirty-nine of these aircraft (AL503-641), plus the 6 LB-30A's and 20 LIB I's scheduled for delivery earlier, were to complete the total of 165 Liberators originally ordered by the British[5]. The Liberator II featured a fuselage

Top: LB-30A AM259 carrying civil registration G-AGCD. /USAF

Above: All 20 Liberator I's (ex-B-24A's) are shown, in various stages of assembly, in this unique 1941 photo. The four aircraft in the bay on the right are even more interesting. They still bear 'LB-30MF' stencilling, indicating they were part of the original French order. However, they have the revised, deeper fuselage and will become the first Liberator II's.

AM92

Above left: AM911 sits for her portrait at San Diego after roll-out in March 1941.

Left: The last of 20 Liberator I's AM929. /USAF

Top: This is 40-702, the sole YB-24 delivered to the USAAF. The Y-designation was later dropped, and it then became the only just-plain-B-24 that ever existed. This aircraft was later converted for transport use at Forth Worth. /USAF

Centre: Front quarter view of the YB-24. /USAF

Above: B-24A's of the Ferrying Command at Bolling Field in October, 1941. No 74, on right, had just returned from Lt Reichers' famous flight to Moscow with the Harriman party. The several aircraft shown here probably pioneered more world-wide air transport routes then any other single type of aircraft. /USAF

stretch ahead of the flight deck and a deepened rear fuselage to provide adequate room for placing a power turret aft of the tail section. The overall fuselage length was thus increased from 63 ft 9 in. to 66 ft 4 in. The aft section change also permitted a better fairing between the fuselage and horizontal stabiliser. A second astro-hatch was provided on top of the fuselage, just ahead of the wing, for use by a 'fighting controller' in directing the defence of the bomber. As delivered, these aircraft had provisions for mounting two Boulton-Paul four-gun power turrets, one in the tail and the other midway along the top of the fuselage over the trailing edge of the wing. The tail turret was the E.Mk.II with servo-feed mechanism and 2,200 (later 2,500) rounds per gun. This unit could rotate 65° on either side of the central position, and guns could be elevated 60° and depressed 50°.

The mid-upper turret was the A.Mk.IV with 600 rounds per gun and spare tanks of similar capacity. Rotation was 360° and elevation 84°, with no depression. Automatic cut-outs protected the tail, propellers and astro hatch. The G.J.3 reflector sight was fitted on both turrets. The tunnel gun feature of the Liberator I was retained, twin gun mounts provided at each waist window, and a new socket-mounted position added in the lower part of the nose compartment. The beam guns also used the G.J.3 sight, while the nose and tunnel guns were fitted with the G.1 prismatic sight. All guns were Browning-Colt ·303 calibre.

The twelve fuel cells and fuel lines were self-sealing, and fed P&W commercial R-1830-S3C4G engines, which later were redesignated R-1830-61's. Propellers were Curtiss Electric. This was unique, for all other Liberator variants before and after the LIB II were fitted with the Hamilton Hydromatic. Including provisions for an increase in bomb load, the Liberator II was initially rated at 46,250 lb gross weight—already some 4 tons heavier than the XB-24.

On 2 June 1941 the first of these aircraft, AL503, took off from Lindbergh Field on its final acceptance flight. As the prototype of its class, AL503 had been fitted with the two Boulton-Paul power turrets at the San Diego factory while the other Liberator II's were to receive this armament after delivery to the UK. During take-off a loose bolt jammed the elevator control in the up position and the aircraft stalled into San Diego Bay, killing Consolidated's chief test pilot William B. Wheatley and all others aboard.[6] In order to fulfil Contract F-677, another Liberator II was built and assigned serial FP685. The last Liberator II was accepted on 6 January 1942.

Beginning on 19 December 1941 the nine B-24C's (40-2378/2386) were rolled out, with the last of this type being delivered on 17 February 1942. The 'C' was designed as the 'production breakdown' version of the B-24 and these aircraft incorporated the turbo-supercharged R-1830-41 engines and self-sealing tanks pioneered on the XB-24B. It was the first USAAF version to be equipped with power-operated turrets, with a Martin unit located just forward of the wing leading edge and a Consolidated turret in the slightly-lengthened tail.

Concurrently with the C's the USAAF received its first B-24D's—which took the place of six of the 'borrowed' aircraft originally ordered as YB-24's and thus carried the early serials 40-696/701.[7] These were accepted between 23 January and 5 February. Equipped with the -43 engine, all were retained in the US for miscellaneous duties except 40-698, which saw service in India and North Africa and was salvaged there on 27 November 1943.[8]

With these aircraft ended what may be considered the initial production period of the Liberator. With the introduction of the distinctive oval-shaped cowling to provide air intakes for intercooler and supercharger, and camouflage paint now standard, the dimensions and outward appearance of the production Liberator would remain relatively unchanged until mid-1943.

The first B-24 visit to the home grounds of its arch rival, the B-17, proved rather embarrassing. In the first place, the Army ordered the trip because it wasn't completely satisfied with Consolidated's flight testing programme. In effect, the Army told CAC to 'go up to Seattle and see how Eddie Allen runs

Top: AL507 and 508 near completion. / General Dynamics

Above: A much-modified BOAC-operated Liberator II, AL592 (G-AHYF).

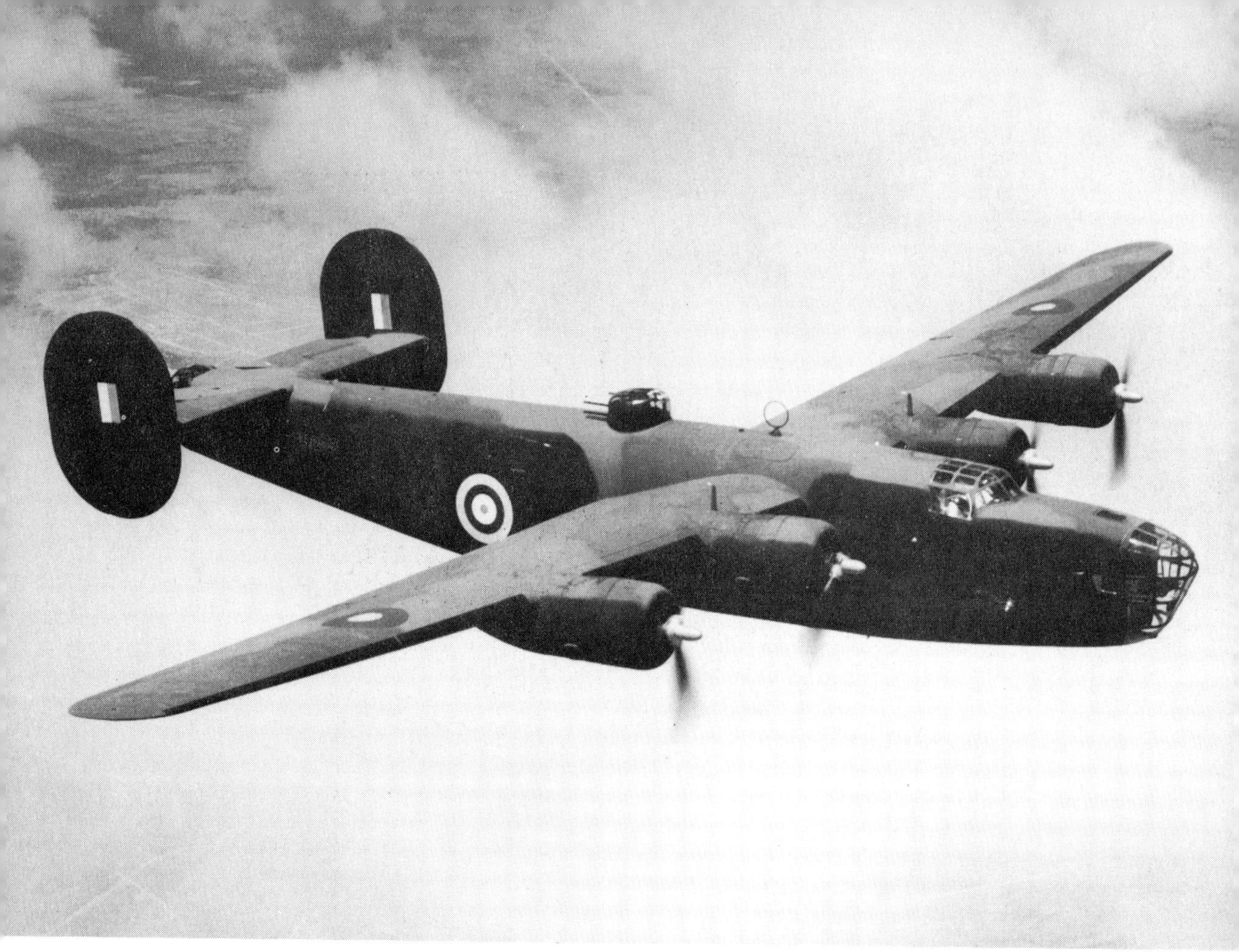

Above: AL503 before disaster struck on 2 June 1941. / Consolidated

Left: William B. Wheatley. The engineer's cap was a trademark. / General Dynamics

Below: First USAAF Liberator to carry a top turret, the B-24C was the direct ancestor of the mass-produced B-24D. Only nine B-24C's were built. The above aircraft, 40-2384, was accepted by the AAF on 12 January 1942. None of the B-24C's saw operational service as bombers.

a *real* flight test department'. CAC flight test engineer Bill Chana recalls: 'On that particular B-24 we were having trouble at the time with one main gear not latching up. Several Boeing people flew the ship and on almost every flight that darn gear would flop down. It was downright embarrassing. When it was time to leave our pilot, Art Bussy, wanted to make sure we departed with dignity so he announced that we would ''hold the gear up with the handle'' just to make sure.'

It didn't work out that way. Just as the gear went up a hydraulic line ruptured, spraying fluid all over the place. Number three engine had to be feathered immediately, down went the landing gear, and the Pride of Consolidated limped out of Seattle with its twin tail between its legs.

Because the name Liberator was first commonly used in association with the early employment of the aircraft by the British, the general impression was formed that the name was British in origin. In fact, this seemed to be confirmed in 1942 when Consolidated ran full-page advertisements in several popular US aviation magazines which stated, in part, 'We always thought of her by her US Army designation—the Consolidated B-24 . . . Then we discovered she had already *won* a name. The British were calling her the Liberator . . . So, from now on, Liberator is official.'

The advertising agency was obviously not in direct touch with R. H. Fleet. On 25 October 1940, Air Commodore B. G. A. Parker of the British Purchasing Commission had written to Major Fleet to ask what name Consolidated had given to the bombers that CAC would soon start delivering to them. In his reply, dated 28 October 1940, Fleet stated the name was 'Consolidated Liberator', adding: 'We chose Liberator because this airplane can carry destruction to the heart of the Hun, and thus help you and us to liberate those nations temporarily finding themselves under Hitler's yoke.'

The 'British origin' forces need not concede total defeat, however. In reply to an inquiry on the subject in 1972, Major Fleet supplied the following additional information. The name Liberator was suggested by his children's governess. Her name was Miss Edith Brocklebank. She was British.[9]

Consolidated's record in bringing their new bomber from drawing board to volume production (by pre-war standards) in only three years was a significant achievement. However, it was not without its price. The design had not been engineered for true mass production, and in fact a great deal of engineering information was incomplete or non-existent. While this shortcoming would not be felt immediately, it was to become a major source of constant trouble in the months—even years—ahead. The Liberator underwent more engineering changes than any other World War II aircraft; San Diego alone catalogued 1,820 of them, or an average of one change for every 3·6 aircraft delivered. Already behind at the start, engineering, as we shall see, never really did catch up. Contributing to the problem was the CAC practice, established over the years, of minimising engineering costs by utilising existing data whenever available. Thus in 1944, when the RY-3 contract was undertaken, the situation was as follows. There was a basic set of master drawings for the B-24. This same set of drawings, less the armament installations and certain structural parts, plus new drawings and engineering change notices, constituted the drawings for the C-87. The basic B-24 drawings, less certain armament, power plant, and structural items, plus new drawings and engineering orders, constituted the drawings for the PB4Y-2. The combination of the two latter sets, plus additional deletions of certain drawings relating to structures and equipment, plus additional drawings and engineering changes, constituted the set of working drawings for the RY-3 . . .

But we're getting ahead of our story.

The B-24D

Fuselage

The Liberator fuselage featured conventional all-metal, semi-monoque, stressed skin construction. Formers were located at stations .1 through 9.2 as shown on the diagram, with those at stations 1.0, 3.0, 4.2,

Top: The second B-24D, 40-697. / USAF

Above: The twelth B-24D, 40-2354, photographed in England, probably in early 1942. Ship has been modified to include a scarf-mounted .50 cal gun and scanning windows in the ventral position. This aircraft was being used as a VIP transport at the time. / Via Roger Freeman

5.1 and 6.0 heavily reinforced as main frames.* The canted former at station 2.0 was also of heavy-duty construction to take the load of the nose wheel gear, which was attached to it in such a fashion that the wheel itself retracted vertically into the fuselage and then nestled rearward. The twin nose wheel doors on all B-24D's opened inward, and featured a quick-release for bombardier-navigator emergency exit. Space inside the fuselage was divided into five major compartments: nose, nose wheel, flight deck, bomb bay and rear. Crew movement between the nose compartment and flight deck was by means of a passageway through the nose wheel compartment around the right side of the nose wheel gear. Movement to the rear of the aircraft was by way of a narrow catwalk along the centreline of the bomb bays. Access to the aircraft from outside was through the nose wheel opening, the entrance hatch in the bottom of the rear fuselage compartment, or through the bomb bay doors. The latter could be opened from the outside by a handle located on the right side of the nose section which activated the bomb door emergency system.

Under normal tyre deflection the fuselage-to-ground clearance was less than 20 in. and this fact, together with the level attitude produced by the tricycle gear, greatly facilitated bomb loading. The twin bomb bays measured 17 ft 10 in. in over-all length. This was nearly twice the size of the single B-17 bay—a fact that Liberator crews invariably pointed out during any discussion on the relative merits of the two aircraft. The only really unusual feature was the flexible bomb bay doors: corrugated sections covered with dural which rolled up along the outside of the fuselage instead of dropping down into the air-stream. These doors, by the way, were invented and patented by a Consolidated employee, Emric Berger, who licenced them to the Government without royalty.

Bomb rack provisions were made for the following bomb loads: 20 x 100 lb, 12 x 300, 500 or 600 lb, 8 x 1,000 or 1,100 lb, 4 x 2,000 lb. Beginning with B-24D-30-CO,[10] stations were provided to carry 8 x 1,600 lb armour-piercing bombs. Bomb bay doors could not be fully closed with this loading, and speed was restricted to 275 mph. Two 4,000 lb bombs could also be carried utilising special demountable underwing racks located between the fuselage and inboard nacelles; however, because it seriously affected flight characteristics this capability was hardly ever used operationally. Centre of gravity considerations made it necessary to always load the front bomb bay first.

The Liberator fuselage had a maximum cross-section height of 10 ft 5 in. and a width of 7 ft 5 in.

Wing

The Liberator wing was probably the most distinctive feature of the aircraft; certainly its high aspect ratio (11·55) was responsible for what modest aesthetic qualities the B-24 had. With an over-all span of an even 110 ft, the wing chord tapered from 14 ft at the root to 5 ft 2 13/32 in. at the edge of the detachable tip sections. The leading edge was swept back at an angle of 3° 30′; wing incidence was 3° and the no-load dihedral angle 3° 26′. A Consolidated Aircraft Corp (CAC) 22% section was used at the root, uniformly reduced to a CAC 9·3% section at the tip. Covering was all metal except for the ailerons. Total wing area, including ailerons but not flaps, was 1,048 sq ft.

Structurally the wing consisted of five major sub-assemblies: the 55 ft 2 in. centre section (permanently attached to the fuselage),[11] two removable outer panels and two detachable tips. Two rather massive main spars were employed, and the Davis wing section was such that these could be located close to the leading and trailing edges, at 10% and 66·2% of chord, respectively. This provided ample room within the structure to accommodate the eighteen (twelve prior to Serial 41-23640) self-sealing fuel cells as well as the large main gear wheel wells.

The centre section structure included two auxiliary plate girder spars, built up of heavy rolled angles and flat sheet rivetted to two of the main wing bulkheads. These spars carried the landing gear loads. All interspar bulk-

* *See page 196.*

Left: A B-24D, 41-1125, sits for her first formal AAF portrait photos. / USAF

Below left: Flight deck controls, B-24D 41-1113. For some reason the airspeed indicator was retouched out of this 1942 photo. / USAF

Right: B-24 interior looking forward from waist window position. / Consolidated

Below: De-mountable Liberator wing rack for carrying 4000 pound bomb. Eglin Field found the idea operationally unsuitable.

heads (ribs) were Wagner beam, truss or pressed sheet, depending on load and access requirements. Ribs were not notched to receive stringers; instead, the latter were bolted to a flange on the rib and the skin was in turn rivetted to the stringers so that the skin did not actually touch the ribs. This type of construction did not inhibit wing flex, a feature easily noticed in many photographs of the Liberator in flight. The entire wing contained sixty-one ribs, including one on the aircraft centreline and three in each tip section. The outer panel splice was located between ribs 13 and 14 at the outer edge of the outboard engine nacelle, with all major attachment bolts accessible through the main landing gear cut-out. Each outer panel weighed 600 lb and each tip 20 lb.

Ailerons were direct-operating (no differential ratio) and aerodynamically balanced. Each had a total area of 41·55 sq ft, an effective area of 31·6 sq ft and 20° movement up or down. A single trim tab of 2·52 sq ft, located near the inboard edge of the right aileron, could be set at plus or minus 10°.

The Fowler-type flaps were built up of aluminium alloy and Alclad sheet, and moved rearward on rollers over five steel tracks attached to lugs on the rear spar. Both flaps were operated by a single hydraulic jack lying along the rear spar on the port side of the aircraft, which extended and retracted the flaps through a system of cables. Maximum extension was 40° down, which gave a 55% increase in lift and a 70% increase in drag. Maximum airspeed at which flaps could be employed was set at 155 mph. Total flap area was 144·1 sq ft, and each flap assembly weighed 125 lb. The flap was designed as a constant chord aerofoil with no 'wash-in'. Since the Davis wing was tapered in planform and section, which resulted in a warped lower surface, it was necessary that the flap also have a warped contour to conform to that of the wing. In early Liberators this warping was accomplished by creating a difference in the rigging tensions of inboard and outboard flap operating cables. Later, landing vibration problems necessitated reinforcement of the flap structure so that warping by cable tension was impossible and a revised inboard stop was provided to reduce the amount of warping required. This caused the inboard trailing edge of the flap to retract to a point slightly aft of its normal position, resulting in a small 'break' in the original uniform taper of the trailing edge of the Liberator wing.

In order to maximise wing effectiveness on the Liberator, paricular design attention was given to engine location. As a result, the upper contour of any nacelle was no more than 1½ in. higher than that part of the top wing surface located behind it. Designer Laddon was particularly proud of this feature.

Two life rafts were stowed under hatches in the fuselage at the point where the front and rear wing spars went through the fuselage. An interior emergency handle released both rafts, which were thrown outward from the aircraft by means of springs while wires attached to CO_2 bottles in the rafts opened valves which automatically inflated the rafts as they were ejected.

Tail

Stabiliser and fins were rivetted Alclad aluminium alloy construction, stressed skin type. Rudders and elevators were of aluminium alloy torque box and rib construction, fabric covered with sheet Alclad leading edges, and were statically and dynamically balanced. The horizontal stabiliser employed an NACA ·0015 aerofoil and had a total area of 124·94 sq ft. The elevators had an effective area (aft of the hinge centre line) of 51·46 sq ft and an area of balance of 15·60 sq ft, for a total area of 67·06 sq ft. Movement was 30° up, 20° down. Two trim tabs (total area 5·95 sq ft[12]) could be set at plus or minus 10°.

The interchangeable vertical fins used an NACA ·0007 flat-sided aerofoil and had a combined area of 122·84 sq ft. Rudders had a total area of 65·0 sq ft, comprised of 48·84 sq ft aft of the hinge centre line and the balance forward of this point. Right or left movement was 20°. The twin tabs (total area 1·92 sq ft) could be set up to 10° right or left. (Total tab area was increased to 3·1 sq ft starting with 42-40753 on CO machines and 42-63926 on CF machines.) Right and left rudders also were

Top: Photographed from certain angles in flight, the Liberator could appear almost graceful. / Consolidated

Centre: Rugged construction was a well-known Liberator hallmark. . . / USAF

Above: This view illustrates Laddon's point about engine location. / US Army

interchangeable by reversing the attaching horns.

The entire Liberator tail assembly was attached to the fuselage with four (later eight) bolts. Some landing-vibration damage was discovered in early aircraft, since no major structural changes had accompanied an increase in span from 24 to 26 ft. However, the assembly was rugged and was eventually stressed to withstand landing load forces of 25G. Before this fix had been accomplished, however, one of the war's strangest Liberator losses occurred. An ATC C-87 (the transport version of the Liberator) approaching Miami, Florida, encountered in-flight tail flutter of such magnitude that its pilot, an experienced veteran, felt it mandatory to abandon the aircraft.[13] After turning the ship to a north-east heading and setting the automatic pilot, he and his passengers bailed out of the bucking Lib. What happened after they left is pure conjecture, but the fact remains that the wreckage of the uninhabited C-87 was later found on the side of a mountain near Monterey, Mexico—1,250 miles due *west* of where it was abandoned.

Landing Gear
The B-24D had a tread of 25 ft 7½ in. The distance between nose and main gear axles (centre line measurement) was 16 ft. The swivelling but non-steerable nose wheel was fitted with a 36 in. x 10-ply tyre while the main gear was 56 in. in diameter and of 16-ply construction. Both were of smooth contour (no tread) design on early B-24D aircraft, but later models were supplied with an over-all diamond tread design. Until B-24D-CO 42-40392 and D-CF 42-63832, a rather complicated emergency nose gear release was standard. This was then abandoned and the following instructions substituted: 'In all subsequent airplanes, nose gear is lowered in an emergency by kicking it out and is independent of the main gear emergency system.'

The final component of the landing gear was the tail bumper, located under the rear fuselage. Very early Liberators (e.g., LB-30, B-24A) used an actual tail wheel for this purpose. This was of a fixed type, permanently mounted with approximately 50% of its diameter outside the aircraft. On the early D models this was changed to a fixed hard rubber bumper type, then later changed again (beginning with 41-23640) to an oleo shock-equipped retractable unit—which was supposed to retract with the landing gear but often failed to do so. In October 1942 all bumper-equipped USAAF Liberators were ordered to be converted to the retractable tail skid.

As a Liberator was taxied to a stop and the brakes applied, it would rock longitudinally and come to rest in either a nose-high or a nose-low attitude due to the packing friction and inertia of the nose wheel shock strut. Normal procedure was to release the brakes and rock the aircraft slightly to level it if the deplaning of the crew had not already accomplished the same purpose.

Engines
All B-24D Liberators used the Pratt & Whitney R-1830-43 engine except the late blocks (from 135 onward) produced at San Diego, which used the -65. These engines differed only in the make of carburettors fitted with the -43 using the Bendix-Stromberg injection type while the -65 used the Chandler-Evans (CECO) direct metering type. These engines were geared 16:9 and were rated at 1,200 hp at sea level. Each turned a Hamilton Standard three-bladed, full-feathering propeller of 11 ft 7 in. diameter. Each engine was attached to its mount by eight flexible shock mounts. The nacelles were designed to permit engine and mount to be removed as a unit. In addition to their regular duties, the No. 1 and No. 2 engines (port outboard and inboard) drove instrument and de-icer system vacuum pumps, while the No. 3 engine drove the main hydraulic system pump.

The distinctive oval cowls of the B-24 resulted from placing air scoops on each side of the engine. Viewed from the front, the right hand opening fed air to the intercooler where it was used to cool the compressed air from the turbo-supercharger, and to the turbine and compressor for direct cooling of these items. The left hand opening provided air for the supercharger itself, the oil cooler, as well as small amounts for cooling the generator and other areas. Each scoop had a small diversion to cool one of the twin magnetos fitted.

All Liberators prior to B-24D 41-23970, plus B-24D 42-63751 through 757 (6 a/c), B-24E 41-28409 through 41-28416 (8 a/c) and B-24E 42-6976 through 42-7005 (30 a/c) had long ring cowlings and used narrow propeller blades. All other Liberators used wide (paddle) blades and had short ring cowls.

Additional data on the Liberator power plant is given in Appendix R.

Systems

The B-24D was basically a hydraulic aeroplane. The main hydraulic system operated the landing gear, wheel brakes, wing flaps and bomb bay doors by pressure taken from the No. 3 engine-driven pump or an auxiliary electric pump. Pressure built up in two accumulators provided parking brake pressure and an emergency source of pressure for operating the bomb bay doors. A hand pump system, located on the floor to the right of the co-pilot's seat, was provided in case both the No. 3 engine and the electric auxiliary pump were put out of action.

The 24 V direct current electrical system was fed by generators on each engine, backed up by two storage batteries and an auxiliary gasoline-driven engine, located in the nose wheel compartment. Although sometimes called by its official acronym APU (for Auxiliary Power Unit), this engine was much more commonly referred to as the 'putt-putt'. This system operated the engine primers and starters, fuel booster pumps, auxiliary hydraulic pump, propeller feathering pumps, cowl flap and intercooler shutter motors, oil dilution solenoids, plus normal instrumentation, lighting and communications equipment.

The main fuel system of the B-24D consisted of twelve self-sealing fuel cells installed in the wing centre section. These were interconnected to form four separate 'tanks', with each tank made up of three interconnected cells. Four booster pumps were installed on the lower wing structure, one directly below each of the four inboard cells. Three interconnected, auxiliary fuel cells were installed in each wing just outboard of the wheel wells beginning with serial 41-23640. Fuel capacities were as follows: tanks 1 and 4, 611 US gallons each; tanks 2 and 3, 571 gallons each; wing tanks, 225 gallons each. In addition, two bomb bay tanks holding 390 gallons each (early type) or 400 gallons (later type) could be installed, giving a maximum fuel capacity of 3,594 (or 3,614) US gallons. These figures are somewhat theoretical, for all self-sealing tanks absorbed a certain amount of gasoline which, in turn, reduced their net capacity. Total oil capacity was 132 gallons.

With wing tanks full, some early Liberators were prone to 'siphon'—i.e., airflow over the top of the wing would create so much negative pressure that gasoline would be sucked out through the small vent holes in the tank filler caps and sometimes around the actual caps themselves. Under certain conditions the flow rate of this effect could reach alarming proportions. Since this raw, 100-octane stream was being released only a few feet above the four turbos mounted on the underside of the engine nacelles, which in turn were apt to be throwing chunks of red hot carbon into the slipstream, the possibility of immediate disappearance via fireball was of considerable concern. Although for some unknown reason certain Liberators were more notorious 'siphoners' than others, it was general practice to limit the filling of all wing tanks to a point some 3 in. below the bottom of the filler necks. In later series B-24's the siphoning problem was largely solved by the introduction of the more efficient 'post type' ram air fuel tank vents which operated on the venturi principle.

Exterior Lights

A retractable, 400 W, sealed beam landing light was located in the under-surface of each wing just inboard of the main wheel well. Navigation lights included one on the top and one on the bottom of each wing tip (port red, starboard green) and one (white) on the

Above: Early B-24D production shot shows 41-1124 *et seq* on the line. /USAF

Below: B-24E production at Fort Worth from Ford Knock-Down kits manufactured at Willow Run.

outboard surface of each vertical stabiliser. There were seven formation lights: three on top of the rear fuselage and four on the horizontal stabiliser. Lenses in all formation lights were blue in colour. A white (later changed to red) passing light was located in the port wing leading edge between the engines. Recognition lights were as follows: one (white) on the top centre of the fuselage, and three (amber, red and green) mounted in the skin under the bomb bay cat-walk. Twin bomb release lights were located at the extreme tail of the Liberator, under the tail turret. A white light went on when the bomb bay doors were fully opened. This white light went out and a red light came on during the bomb release period and for five seconds thereafter.

Antenna System
The typical B-24D left the factory with six antennae, the purpose and location of which were as follows: Command antenna, a single wire extending aft from alongside the top turret to the top of the left vertical stabiliser; Liaison fixed antenna, identical to the Command antenna but on the right side of the aircraft; Radio Compass antenna, a vertical 'whip' mounted on top of the fuselage at Station 5.1; Marker Beacon antenna, mounted beneath the bomb bay cat-walk on stand-off insulators; Radio Compass loop, enclosed in a tear drop housing on top of the fuselage between Stations 5.3 and 5.4; and the Liaison trailing antenna, a single 200 ft wire, weighted on the end and controlled by an electrically operated reel located under the flight deck.

This, then, was the product that Consolidated had to offer. And it was a seller's market if ever there was one.

Multi-Plant Production
In the autumn of 1940, over a year before Pearl Harbor, the Government began to look to the automobile industry to augment aircraft production capacity. Quite understandably this move raised fears among the aircraft companies that such emergency collaboration might result in post-emergency competition. In light of this, it was decided that the auto industry would be brought in not as a plane-maker, but as a parts-maker. With one major exception this decision went uncontested throughout World War II. That exception was the Ford Motor Company. The Ford story, one of the most fascinating aspects of Liberator production, will be described in some detail in a later chapter.

In order to meet the projected large requirements for the B-24, a Liberator production pool had been established in early 1941. The programme called for Consolidated to set up a second assembly plant at Fort Worth, Texas, to supplement the production of the home factory at San Diego, and for Douglas Aircraft to operate a similar assembly plant at Tulsa, Oklahoma. These Government-financed plants would assemble finished B-24's from knock-down or 'KD' assemblies supplied by the Ford Motor Company from another new plant to be built at Willow Run, Michigan. Each KD, which Ford initially contracted to supply at the rate of 100 per month, included all major components of the bomber—fuselage, wings, tail and landing gear. First production at the new plants was scheduled for early 1942. Somewhat later (January 1942) a fourth company was brought into the picture when North American Aviation was given a letter of intent (preceding a contract issued on 1 May) to operate a complete and independent B-24 fabrication assembly complex at Dallas, Texas. It has been noted that the pool arrangement placed Douglas in an unenvied situation—the company was responsible for a Consolidated product whose parts had been fabricated by Ford. Douglas had no authority or control over design, material procurement, detailed inspection, tooling, or assembly methods—yet Douglas was responsible for on-schedule delivery of the finished product.

Thus the overall production plan called for Liberators to be fabricated at San Diego, Willow Run and Dallas, and assembled at San Diego, Fort Worth, Dallas and Tulsa. From the beginning, however, Ford had demonstrated great interest in becoming a full-fledged producer of complete aircraft—as

Above: Prior to the start of assembly operations, the Fort Worth plant was used as a modification centre for aircraft produced at San Diego. This explains why some photos, such as the above, show B-24D-CO machines apparently being built at Fort Worth. / Consolidated

Below: The first Liberator from Tulsa, B-24D-DT 41-11754, was rolled out in July 1942. Note aircraft does not yet have armament installed. Components for this aircraft were shipped from San Diego for assembly at Tulsa and to serve as a 'practice airplane' to orient Douglas plant workers. Number '259' on fin was probably ficticious serial applied for publicity photo. / Douglas

evidenced by the huge Willow Run facility which, although ostensibly built as a parts plant, was in fact an all-out production layout as later events would prove. (In less than three years Ford built 6,792 B-24's, furnished 1,893 others in KD form to Tulsa and Fort Worth, produced mountains of spare parts, and never *did* operate Willow Run at full capacity.) In October 1941 Ford's desire was formally recognised by a contract under which the Detroit firm would turn out completed Liberators in addition to the KD assemblies for Tulsa and Fort Worth. At the same time the scheduled KD production rate was increased to 150 per month.

All this could not be accomplished quite as quickly as original plans had called for and as a result the Tulsa and Fort Worth plants were ready to be fed before Willow Run could feed them. To fill in, both Douglas and Consolidated-Fort Worth began operations with sub-assemblies supplied by San Diego. The aircraft resulting from this temporary arrangement were for all practical purposes identical with the B-24D-CO machines then being built at San Diego, and bore serial numbers originally allocated to the California plant.

The USAAF accepted the first Willow Run Liberator from Ford on 1 September 1942. Designated B-24E to differentiate it from the Consolidated B-24D from which it was literally copied, it was followed by two dozen more before the end of the year. The first KD from Willow Run arrived at Tulsa during September 1942 and by February 1943 Douglas was in series production of the B-24E. Fort Worth received its initial Ford KD in January 1943 and, starting in July, began delivering assembled Willow Run E's in addition to its B-24D production.

At first the Ford organisation had a difficult time in applying the mass-production techniques of the car industry to the production of aircraft. They found, for instance, that they had to re-do the 30,000 drawings they had copied from Consolidated so that relatively unskilled workmen could interpret them. They discovered that working with aluminium was different from working with steel—the two metals just didn't behave in the same way. Worst of all, they had to learn to accept the constant design changes that characterised wartime aircraft production as contrasted to the yearly design freeze of the pre-war car industry.

In the end Ford did master these problems and many more, but the time it took to do so rendered the entire B-24E production run obsolescent almost before the aircraft were completed. As a result, these aircraft were placed in the 'limited standard' service category on 18 September 1943 (the very day the last E was delivered) and practically none reached overseas combat operations. Some E's did make it, however, and *as late as 20 June 1944* the 448th Bomb Group lost B-24E-10-CF 41-29013 on a deep penetration beyond Berlin to Stettin, Poland.[14]

The manufacture of knock-down kits at Willow Run for final assembly by Consolidated-Fort Worth and Douglas-Tulsa meant that partially assembled aircraft had to be shipped overland 950 miles to Oklahoma and 1,250 miles to Texas. For this purpose principal dependence was placed on a fleet of eighty-six specially designed trailers, although limited shipments were made by rail to guard against disruption of the truck schedule because of weather conditions or accidents. Two experimental trailers were constructed and their feasibility proved in trial runs. Sixty-four were then ordered and later twenty more were added.

A knock-down kit consisted of three large units—the aft fuselage section, the nose or forward section, and the centre wing section—plus a variety of smaller assemblies. Shipment of one aeroplane required four railway wagons or two and one-half trailer loads. Under ideal conditions trips to the assembly plants were made in four days, with the usual time ranging from four to seven. Drivers worked four days out of seven, and while on the road they worked twenty-four hours a day. Two drivers were assigned to each trailer and they changed over at five-hour intervals. A bed was built into the tractor unit.

The trailers were 63 ft long, 12½ ft high and 8 ft wide. The overall length of the trailer and tractor was 75 ft. Dimensions of all the

Above: Although its serial is higher than some later deliveries, this is the first Liberator to be assembled at Fort Worth. Ship was built from sub-assemblies shipped from parent plant at San Diego.

Below: B-24D-4-CF's, destined for USAAF anti-sub use, on the Fort Worth line during June 1943. / Consolidated

Above: Sub-assemblies ready for rail shipment to Fort Worth. These nose section are San Diego C/n's 84 and 85, which were assigned USAAF serials 41-11588 and 41-11589. /General Dynamics

Below: B-24E-10-DT's on the Douglas assembly line at Tulsa. Note easy access to equipment in wing via the hinged leading edge. Spring-loaded life raft containers are in open position on first aircraft. SBD's in background. /Douglas

trailers were the same, but the interiors differed. One type was used to carry centre wings, outer wings and other related assemblies; another to carry fuselage sections, both forward and aft; while the third was used to carry spare parts and purchase order items. The top of the trailer consisted of five sections of removable framework, covered with waterproof canvas. When loading or unloading the canvas and sections of framework were removed.[15]

The tractor units for the trailers weighed over 4½ tons, and were powered by two engines. The unit had two rear axles, each axle being driven independently. The motors were located side by side and were equipped with a fuel and oiling system that permitted the trips to be made non-stop.

When empty, the trailers and tractors weighed about 21,000 lb or 10½ tons. The fuselage trailer carried a load of about 8,000 lb and the wing trailer twice as much or about 16,000 lb. The maximum load for a trailer was 27,000 lb or about 14 tons.

The fleet of eighty-six trailers averaged a total of 250 round trips a month, with the record number of round trips in one month being 375. The weather proved to be the only serious operating hazard, and occasional floods or heavy snowfall would tie up as many as six to ten trailers at points along the route.

Production at North American began slowly. The first B-24G, also basically a D aeroplane but with another new designation to denote its separate origin, was flight tested in January 1943 and accepted in March. By the end of July only four more B-24G's had been accepted by the USAAF and it was not until January 1944 that production at Dallas passed the one-a-day mark.

By 30 June 1943 the USAAF had accepted an all-model total of 2,988 Liberators. On the same date, 1,867 of these were still available in the inventory of the USAAF for the job ahead and hundreds of others flew in US Navy and RAF markings. But there were thousands more to come—more than Consolidated, David Davis or the US Army Air Corps could have imagined back in 1938. The numbers under contract grew so large, in fact, that with effect from 1 April 1943 a new agreement had been negotiated with Mr Davis whereby his royalty payment per Liberator was reduced from $92·71 (the going rate under the 1/16 of 1% feature of his original agreement with CAC) to a straight $5·00 per aeroplane. He received this amount for each of 17,383 Davis-winged aircraft produced on and after that date.

But although the number of Liberators to be built grew larger, the days of the D and E on the production line were numbered. On 30 June 1943 the first of a new breed—the B-24H—was delivered from the Ford assembly line at Willow Run, and 18 September 1943 saw the last of the early Liberators leave the factory.

At this point it may be well to pause for a moment and go into the subject of Liberator armament in some detail, inasmuch as this will be discussed with increasing frequency in the sections that follow and, in the past, has often been the source of considerable confusion. It should be noted, however, that the comments which follow cover factory production configurations and should not be assumed to deal with operational theatre modifications, which abounded. (For example, the B-24D's of the HALPRO Mission were given two fixed, forward firing ·50's in the lower nose compartment. These were aimed by the pilot who had, mounted outside the windscreen on the top of the fuselage nose, an honest-to-goodness ring and bead site for the purpose.)

The very first D-CO's (through 41-1142) had only one ball-and-socket mounted ·50 calibre machine gun in the nose to supplement the twin ·50's in the top and tail turrets. Beginning with 41-11587 the Bendix lower turret was installed. This electrically operated and retractable unit, which was aimed through a periscope by a gunner kneeling inside the fuselage, was never a satisfactory arrangement and the last Liberator to leave San Diego with this installation was 41-11874. To replace it, the old tunnel gun of the Liberator I was resurrected. This employed a single ·50 calibre gun, socket-mounted at the centre of a hinged

Above: B-24E's at Willow Run, 1943.

Below: A rare bird — the first of the initial 25 B-24G's, which were completed without nose turrets. Another aircraft from this production batch, 42-78056, was tested at the AAF Proving Ground at Eglin Field during October 1943 and found to be tail heavy under all load conditions—a situation helped somewhat by the installation of the Emerson turret in the nose of later B-24G's. /North American

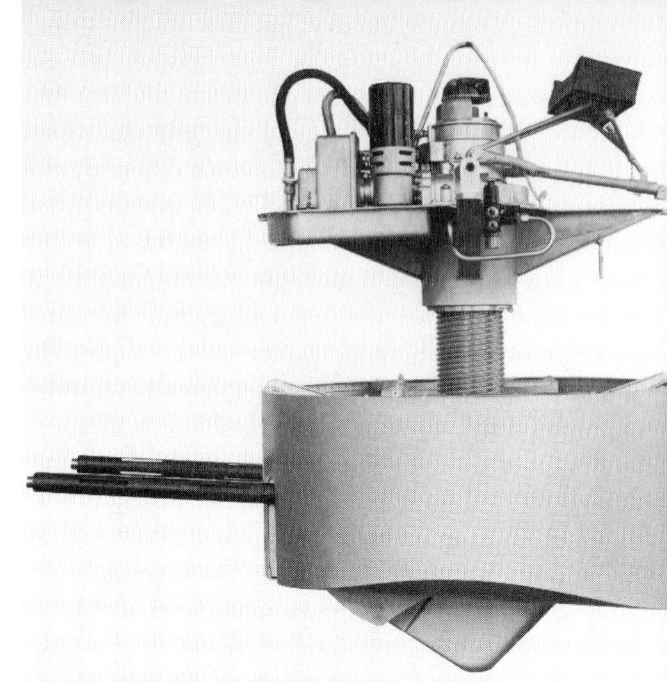

Right: The Bendix lower turret. / Bendix Corp

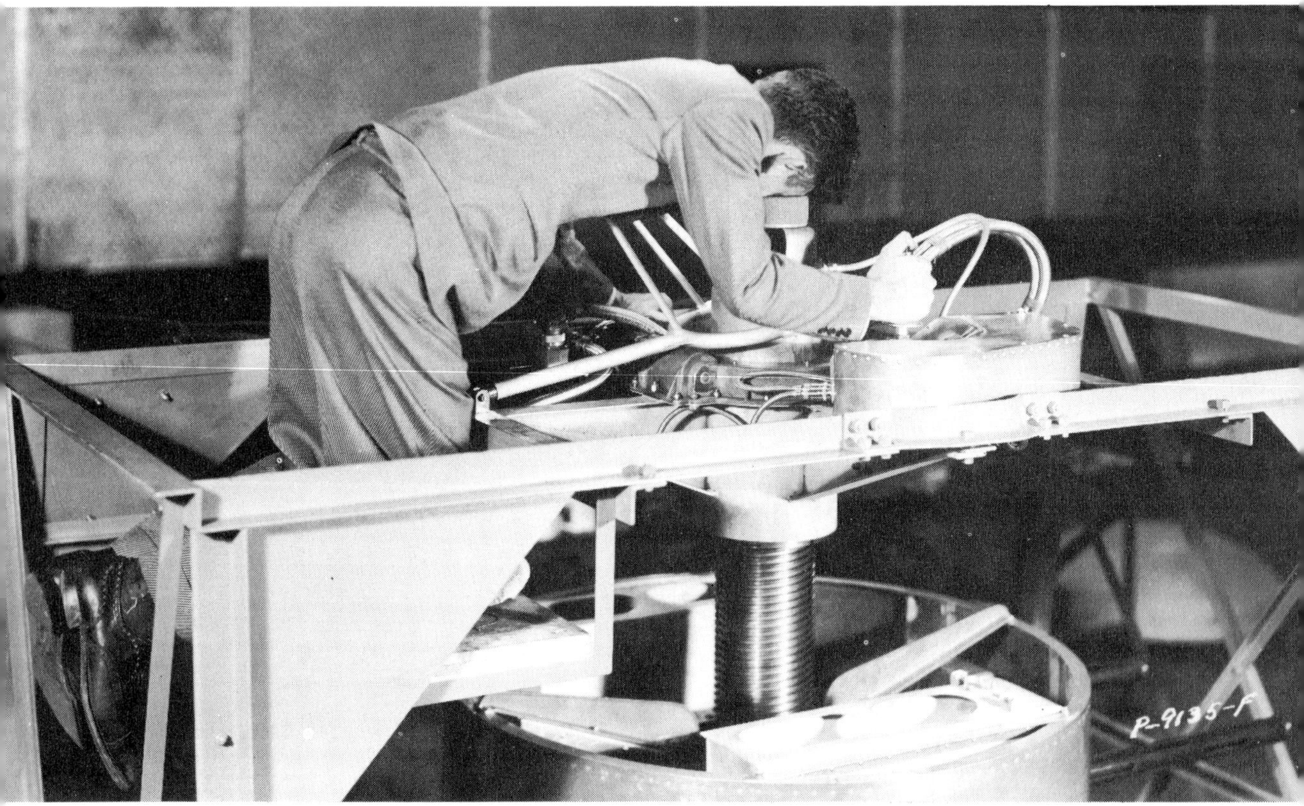

Below: Mockup installation showing gunner's position. / Bendix Corp

frame which was swung down into the rear hatch opening when in use. Small 'tunnel gun scanning windows' were added to the lower rear fuselage to increase visibility for the gunner. By Tech Order, a field change was instituted to remove all Bendix lower turrets then in service and install tunnel guns. The turret opening was then faired over.

The final B-24D-CO's, beginning with 42-41164, eliminated the tunnel gun in favour of the Briggs/Sperry (Sperry design, manufactured by Briggs) ball turret with its twin ·50's. It is interesting to note, however, that the vestigial tunnel gun scanning windows remained on San Diego machines until the B-24J-160-CO.

The single socket-mounted nose gun was augmented by two more through the greenhouse beginning with 41-24220, and at the same time a flexible ·50 was installed at each waist opening. This configuration continued through 42-40058, when two of the nose guns were moved aft of the nose enclosure and small sighting windows provided.

All Fort Worth-manufactured B-24D's, besides the standard top and tail turrets, were of the three-in-the-nose, two-in-the-waist, and one-in-the-tunnel variety.

Willow Run B-24E's had the single nose gun, top and tail turrets through 42-7005. Beginning with Block 5 the tunnel and waist guns were added, and at 42-7066 the two additional nose guns appeared. The Bendix turret was never fitted to the E, although structural provision for it was included through the end of Block 10.

The initial twenty-five G-NT's all had the three ·50's up front. Of these G's, however, the first four had no belly armament at all, the fifth had a tunnel gun, while the rest had the lower ball.

The original B-24 power tail turret was the CAC 5800-2, or A-6, model. This unit, which was installed on all B-24D's through serial 42-40157, was then replaced by the 5800-3 or A-6A, which for the first time increased gunner visibility by means of transparent panels on the sides. The guns in these units were mounted assymetrically, so that the left gun barrel protruded about 6 in. farther than the right. Because of interior fittings the side transparent panels were also of different size and shape, with the right side being considerably smaller. A-6A installation continued on CO aircraft through serial 42-100200, on CF aircraft through 42-64306, and was original equipment on machines assembled by Ford, Douglas, Fort Worth and North American through block H-1 and G-1, respectively.

The A-6A was superseded by the 5800-5 or A-6B, which featured symmetrical gun mounting and considerably larger transparent areas on both sides (although the right side was still somewhat smaller than the left). The A-6B was manufactured by the Motor Products Corporation (MPC), and remained standard for the balance of all B-24H and B-24J production except at Douglas, where supply problems dictated a return to the A-6A at the start of block H-25. Douglas continued to install this version until the end of DT production.

With some additional fairings provided for aero-dynamic reasons, the A-6A and A-6B also provided the inventory of B-24 nose turrets until the universal adoption of the Emerson A-15 type.[16] The A-6A was used on nose-turreted CO machines through 42-100155 and on CF B-24's through 42-64338. The A-6B then took over until the advent of the A-15 at J-185-CO and J-45-CF. All A-6 series turrets were driven hydraulically by pressure built up by a hydraulic pump, which in turn was driven by a constant speed electric motor. Guns could be moved 75° in azimuth on either side of the aircraft centre line, elevated 71° and depressed to -45°. Armour consisted of a 2 1/8 in. 'bulletproof' glass panel directly in front of the gunner, a 7/8 in. armour plate panel directly below and a 3/8 in. armour plate on each side. Entry to the A-6 turrets was through two sets of doors, one pair in the fuselage and the other pair in the rear of the turret. The latter was often removed in service to permit the wearing of a back pack parachute by the gunner and to generally improve the chances for a successful emergency exit.

The Emerson A-15, which replaced the A-6 in the nose of all later-series USAAF Lib-

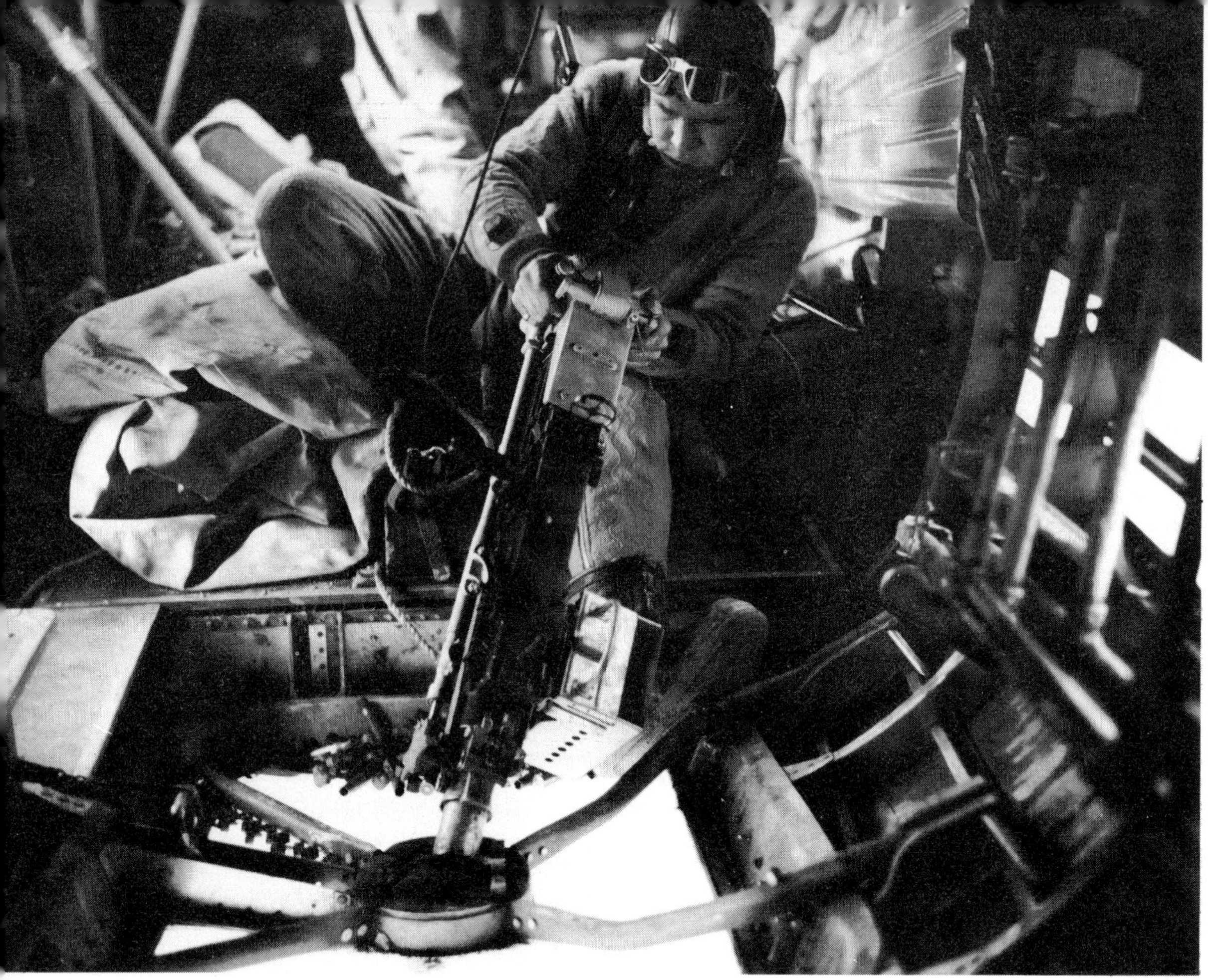

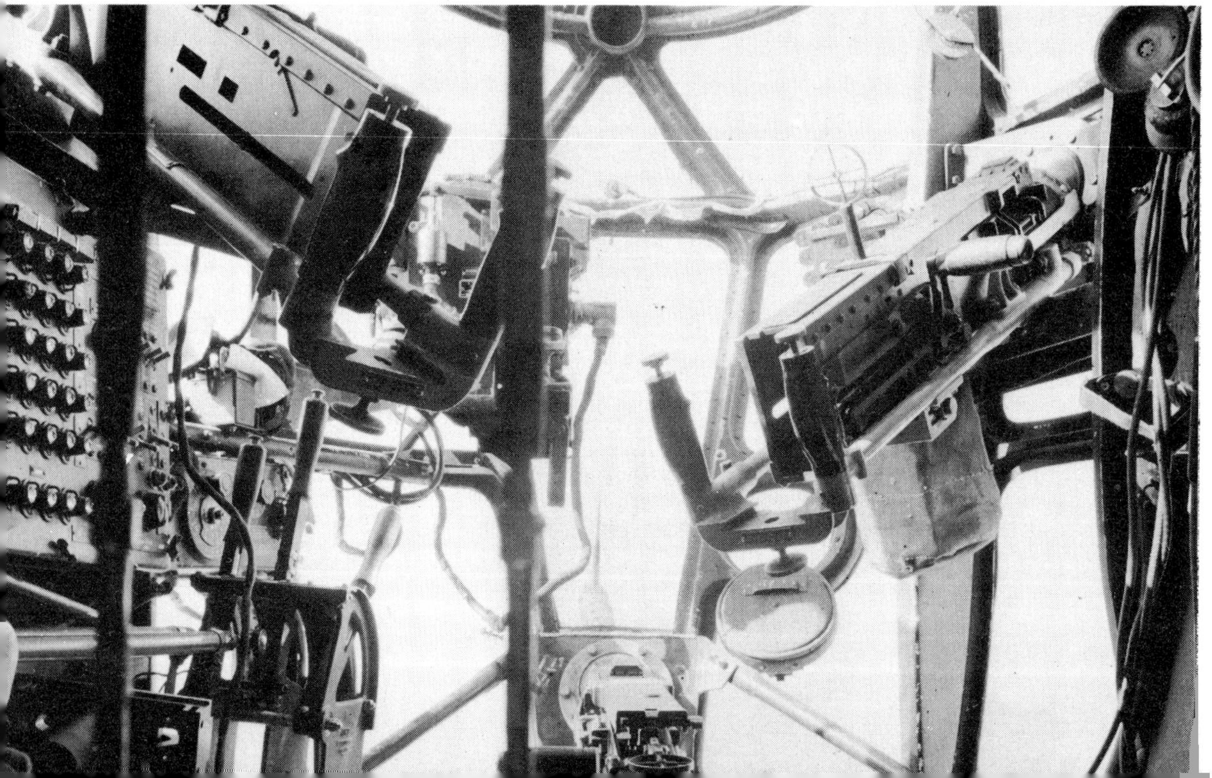

Left: The B-24 tunnel gun in firing position. /US Army

Below left: Three socket-mounted fifties in the nose got a little crowded.

Above: The transparent tops of Liberator turrets were inscribed with degree lines which measured the orientation of the turret relative to the aircraft. The gunner, by glancing upward, could read this information against a pointer which was fixed to the fuselage and jutted out over the top of the turret. The front transparency was similarly marked to show the number of degrees of gun elevation.

Above: The MPC A-6B tail turret in a B-24J-CO. / General Dynamics

Below: This B-24E, 42-6981, has been fitted with a Bendix chin turret. Origin of this modification is unknown, but see footnote. /USAF

erators, was an adaptation of the Emerson Model 111. Design of this unit was originated in early 1942 as a replacement for the CAC tail turret. However, shortly after the initial test model had been received with enthusiasm by the USAAF, more urgent priority was given to the efforts to provide a nose turret on the B-24. The Emerson 111 design, revised on a crash basis and re-christened the Model 127, was pressed into service. Emerson engineers, working together with CAC personnel at Consolidated's Tucson Modification Center, designed the nose installation at the same time the first Model 111 turret was being modified at Emerson's St Louis plant. The initial test unit was completed in early 1943, and following Eglin Field tests was approved for service use as the Army Type A-15. Meanwhile, a considerable quantity of Model 111 tail turrets had been shipped to Willow Run where, as noted later in the text, they were hurriedly converted to A-15 configuration to meet the deadline for the appearance of the B-24H.

The A-15 covered 75° to right or left in azimuth, and +60° to -50° in elevation. Electrically driven, it could be operated at two speeds—normal tracking and high speed. Later models used cast magnesium for all principal structural items as a weight-saving feature.

The continued use of the A-6B nose turret on CO and CF machines after the debut of the A-15 was strictly a matter of supply, and, in fact, if sufficient numbers had been available the A-15 probably would also have gone as well into the Liberator tail as originally intended.

The Briggs/Sperry ball turret used on the B-24 was produced in only two basic versions, the A-13 and A-13A. These were quite similar in appearance and are not very useful references for Liberator identification. The first production line installation of the A-13 was on B-24D-CO serial 42-41164, and this version was not supplanted by the A-13A until block J-15-FO at Ford and block J-90-CF at Fort Worth. San Diego continued to install the A-13 until the end of block M-15-CO, while Douglas and North American used the earlier mark throughout their production runs. Most B-24 ball turrets were lowered and raised by means of a hand-operated winch located on the overhead centreline of the waist compartment just forward of the ball turret opening. Later versions utilised a hydraulic system which was activated by means of a hand pump mounted on the fuselage wall. On these aircraft, emergency raising of the turret was accomplished by means of the Liberator's built-in bomb hoists.

Getting into the ball was a careful operation, for unless locked securely the heavy and delicately-balanced unit could swivel and break a man's leg or snap him nearly in two as he attempted to enter.

The turret rotated 360° in azimuth and the guns could be depressed from 0° to a full -90°, the latter being the entry/exit position. An armour plate panel formed the bottom of the seat and extended up the gunner's back as far as the door hinge. Although fortunately a rare necessity, the Liberator *could* be landed with the ball extended; the usual procedure was to concentrate as much crew weight as possible in the forward part of the aircraft, touch down all wheels simultaneously, and apply firm and steady braking as soon as possible.

The Martin upper turret, like the Emerson driven electrically, also featured both normal tracking and high speed operation. Elevation was from -6½° to +85°, and the unit rotated 360° in azimuth. Armour plate of 3/8 in. (later ½ in.) thickness protected the gunner, although this shield rotated with the turret and therefore was effective only against frontal fire. Interrupters prevented accidental damage to the aircraft tail assembly; however, no similar protection was provided for the propellers arcs and extreme caution was the rule during level firing to port or starboard quarter!

The dependable 'Martin upper' underwent few changes during its wartime life with the B-24 and PB4Y-2 Privateer, the main revisions being associated almost wholly with the type of guns installed. The A-3B featured direct manual charging, while the A-3C and A-3D employed manual charging cables. The Plexiglas enclosure also changed, growing successively larger with each revision.

Above: B-24 number 41-24100, the first D-20-CO, with nose turret added. This Hawaiian Air Depot (7th Air Force) installation did not increase the depth of the fuselage to produce the 'lower lip' effect of later Liberators with factory-installed nose turrets. As a result, room for the bombardier was extremely limited. / USAF

Below: B-24H-25-FO 42-95051. Waist window size has not yet been increased. / Ford

A majority of PB4Y-1's and PB4Y-2's (Privateers) were fitted with the Erco 250 SH (for Spherical Hydraulic) nose turret. This unit could be rotated 85° to either side, elevated 82° and depressed 83°. PB4Y-2's also employed the Erco 250 (or 260) TH (Tear-drop Hydraulic) at the waist positions. The two ·50 calibre Brownings in this turret could be swung 79° aft, 56° forward, 60° up and 95° down, with fire interrupters provided to protect wing and horizontal stabiliser.

The B-24H

To recap, by mid-summer of 1943 Liberator production was under way at all five locations. Consolidated, now CVAC,[17] was producing B-24D's at San Diego, California, at the rate of 200 per month and assembling B-24E's at Fort Worth, Texas, at better than half this rate from KD assemblies supplied by Ford from Willow Run. Ford was also supplying B-24E's in KD form to Tulsa for assembly by Douglas, as well as assembling complete aircraft on its own at Willow Run. Production at Dallas by the newest addition to the Liberator Production Pool, North American Aviation, was just getting under way on the B-24G.

All of these models—the B-24D, E and initial version of the G—were essentially the same aeroplane. However, something new was brewing at Willow Run and on 30 June 1943 Ford delivered the first B-24H, serial 42-7465. Although the honour has usually been credited to the B-24G-1-NT, the Ford H was the first production Liberator to have a nose turret. However, the word 'production' should be emphasised, for nose-turreted B-24's were already operating in the Pacific in considerable numbers. These were initially field modifications utilising spare or salvaged Consolidated tail turrets, but later a full-scale theatre modification programme was undertaken by the Hawaiian Air Depot to modify substantial numbers of D's to the new nose configuration. All of these used the Consolidated A-6 turret.

The advent of the B-24H marked Ford's entry into the real world of producing combat-qualified Liberators. Known at Willow Run as the '801 change' (because it was to become effective with Ford's 801st aeroplane, counting KD's) the B-24H revision not only incorporated some fifty-six separate master changes but was carried out at a frantic pace inasmuch as it was requested by the USAAF in mid-March with a goal of fifty of the new aircraft to be available by 30 June.

Because of the short time schedule, engineering data from Consolidated was necessarily incomplete. In order to modify the nose to accommodate the turret, for example, Ford worked from sketches made at the Modification Center where Consolidated and Emerson personnel were developing the change. Using the sketches, Ford built a plywood mock-up, from the plywood mock-up made a plaster mock-up, and then used the latter to design the new tooling required. The situation was further complicated by the fact that the first Emerson turrets shipped to Willow Run had been designed for use as tail turrets and had to be re-worked before they could be installed in the nose. The original CVAC design for the new downward-opening nose wheel doors was found to be unsatisfactory and the doors had to be redesigned by Ford, while the installation of the first ball turret by Willow Run during the same '801 change' presented additional difficulties to overcome and lessons to be learned.

All in all, it is not surprising that Ford did not meet the USAAF request for fifty H's by 30 June. They did, however, deliver six on that date and forty-two in July—with many of the new parts made by hand because neither Ford nor outside suppliers had had time to tool up for the changes. Practically all of these early H-FO's (Ford retained the first one, 42-7465) were immediately sent into combat in Europe as original equipment for the 392nd Bomb Group.

The Emerson A-15 turret used in the Ford H was some 190 lb heavier than the Consolidated A-6 and had a slightly smaller field of fire. However, it was rated superior because it was electrically powered and more responsive than the hydraulically operated CAC design. The addition of the Emerson increased the over-all length of the B-24H to 67 ft 3 3/16 in., or about 1 ft longer than the D or E. In fitting the

A Ford B-24H. Twin aileron trim tabs and staggered waist guns clearly visible. /USAF

nose turret, Ford used a wrap-around Alclad panel that met the skin of the fuselage in a gentle S-curve. Although giving somewhat of a 'tacked-on' look, this particular feature was unique to the H and J assemblies produced by Ford, and is an infallible recognition feature of these aircraft.

As well as the Emerson unit in the nose, the B-24-H-FO carried the Briggs A-13 44 in. retractable ball turret first introduced during block B-24D-140-CO, the Martin A-3C electric top turret and the CAC/Motor Products A-6A tail turret. (This was changed to the A-6B at the start of Block 5.) One flexible ·50 was carried at each waist window position which, when opened on combat missions, was unprotected except for an adjustable wind deflector on the outside of the fuselage just forward of the opening. This open combat window feature had been standard on all Liberators up to this time but operational reports, particularly from Europe, stressed the high rate of frostbite among waist gunners that was resulting from long exposure at high altitude. To improve the situation, trial modifications were undertaken by Ford and by Northwest Airlines' St Paul Modification Center. While a few of the three-sectioned 'bay window' St Paul installations did reach operational service, it was the Ford solution that was accepted for incorporation in production models. This version enclosed the entire opening with a single Plexiglas sheet fitted flush with the skin of the fuselage, with the gun mounted in a universal swivel below the base of the window. This was the so-called K-6 mount. The outside wind deflector was eliminated. To lessen interference between the two waist gunners the starboard mount was positioned towards the front of the window and the port mount towards the rear. Because of the Ford-Fort Worth-Douglas production tie-in, this modification took place during Block H-20 at each location, although (because of the time consumed by shipment of sub-assemblies) the same block was always being assembled somewhat later at Fort Worth and Tulsa than at Willow Run. In later blocks the size of the Plexiglas-covered area was increased by dropping the sill of the waist opening so that it was even with the bottom of the gun swivel.

All Fort Worth and Tulsa H's also used the Emerson nose turret, switched to the A-6B tail turret at Block 5, the K-6 waist mount at Block 20, and at Block H-25 all three assembly lines went over to a revised Martin upper turret (the A-3D or 'high hat') which featured a raised contour in back and a more sloping front. Block H-25 at all sites also saw deletion of the tunnel gun scanning windows in the lower rear fuselage, while at the same time Douglas alone returned to the A-6A tail turret. Since paint was applied at the site of assembly, its deletion was a function of time (March 1944) rather than block numbers. Thus H-FO's appeared in natural metal finish (NMF) beginning with Block 25; H-DT's and H-CF's beginning with Block H-15.

The B-24H was probably the most popular mass-produced bomber version of the Liberator as far as crews were concerned. Used extensively in Europe, it possessed the all-important nose turret lacking on the earlier D's, yet was generally lighter than its contemporary, the B-24J, and a good pilot could coax it to an additional 1,000 ft or so of altitude. Another highly desirable feature of the H was the addition of a tab on the left aileron. On the Liberator, up-tab deflection had less aerodynamic effect than down-tab deflection due in large part to certain inherent characteristics of the Davis aerofoil. After 4-5° of up travel the efficiency of the tab dropped off rapidly, and extreme up-tab deflection was necessary to correct slight amounts of opposite wing heaviness. The addition of a tab on the left aileron made it possible to trim either wing by use of the more efficient down-tab correction. Why Consolidated waited until the B-24L-5-CO to incorporate this feature is a complete mystery.

The 392nd Bomb Group in particular made an active if unofficial effort to standardise on the H and were well-known for their willingness to swap any brand-new B-24J replacement aircraft they received for 'one used B-24H in good condition'. This in spite of the preference of most pilots for the C-1 autopilot over the H's A-5 model. One feature that did not meet with European theatre approval, however, was the Ford 'coffin seat'—a shroud of armour that afforded good protection but prevented the wearing of a parachute and was a considerable hindrance to a fast emergency exit. These seats were normally removed in service in Europe and protection provided by a 'flak curtain' across the rear of the flight deck.

By 31 December 1943 the total B-24 production programme called for 20,353 bombers, of which 9,104 were to be of the H version.[18] Less than three months later, Willow Run was turning out completed B-24H's at an average rate of better than one every 100 minutes around the clock, seven days a week. And this period included a Block change, from Block 25 to Block 30, with a dozen associated 'master changes'. The plant could have gone to 600 a month but, in truth, the rate of supply had overrun the ability of the USAAF to assimilate it. As historians would later admit, the long rows of completed but unused bombers that began to accumulate outside the assembly plants became a matter of acute embarrassment to the Air Force. A far cry from the dark days of 1941-2!

The B-24G

Meanwhile North American had finished their twenty-fifth B-24G-NT at Dallas during October 1943 and at that point began installing the Emerson nose turret at the start of Block G-1-NT. Other armament was the same as the B-24H-1 through -15-FO. The first of these aircraft was delivered on 3 November 1943. Somewhat later the North American production line pioneered a change particularly welcome to Liberator pilots. This involved the substitution of bell-cranks in place of the gear box in the aileron control system to provide a single-point system of aileron control, and reduced considerably the stick forces required to fly the aeroplane. First introduced on the B-24G-10-NT in January 1944, the change was picked up by Consolidated on the J-30-CF in March and the L-5-CO in September, while Ford began using the revised system in October on the L-15-FO. For some reason all B-24's produced by North American deleted the provision for underwing external bomb stowage even though the structural reinforcement necessary for this capability was built into the aircraft.

Towards the end of the B-24G production run, NAA was tooling up to introduce the first heated-surface type of anti-icing equipment, which would replace the rubber boot type common to all Liberators since the early models. The advent of the changeover to the B-24J-NT, however, made it necessary to postpone the implementation of the de-icer change. Such re-ordering of the priorities of pending changes was the rule, rather than the exception, at all of the production sites and in the final analysis the hot air units never appeared on Dallas-built B-24's. San Diego eventually converted to the heated-surface de-icers with the J-150-CO, and Willow Run followed on the M-20-FO. Fort Worth and Tulsa, like Dallas, never turned out Liberators with this feature.

The B-24G represented a wholly North American operation—that is, as far as production was concerned the NAA Dallas plant was independent of the other members of the production pool—and the initial G's were inclined to be heavier than the H or J. As production continued, however, NAA were able to take steps to lighten their product so that the later B-24G-NT examples did not suffer from this handicap.

All G's after Block 10 were delivered in natural metal finish. A very high percentage of the 430 B-24G's produced saw action in the Mediterranean, while the Eighth Air Force in England was not supplied with any of this series.

The B-24J

The J was built in greater numbers than any other Liberator, and was the only version produced by all five member plants of the Production Pool. Beginning with the San Diego Plant, which went directly from D to J production on 31 August 1943, J's were started at Fort Worth on 25 September 1943, at Willow Run on 18 April 1944 and at Dallas and Tulsa on 14 May 1944.

As with the change from E-FO to H-FO, the

Top: B-24G-15-NT 42-78349. / NASM

Above: A B-24J-25-CO profile in all its classic beauty! / Consolidated

significant difference between the D-CO and J-CO was the addition of the nose turret. The J-CO, however, used the Consolidated-type A-6A, which was almost identical with the standard Liberator tail turret. The familiar sloping front of this unit added a few inches more than did the Emerson, making this the longest series-produced USAAF Liberator at 67 ft 7 5/8 in. Other defensive armament, which did not change until Block 70, included the Martin A-3C upper turret, Briggs A-13 ball, A-6A tail turret and flexible ·50's in open waist windows. During Block 70 there started a series of turret changes which are recapped in the following table:

	Nose	*Upper*	*Lower*	*Waist*	*Tail*
J-1-CO through J-70-CO 42-100124	A6A	A3C	A-13	Flex	A6A
J-70-CO 42-100125 to J-75-CO 42-100156	A6A	A3B	A-13	Flex	A6A
J-75-CO 42-100156 to J-80-CO 42-100200	A6B	A3B	A-13	Flex	A6A
J-80-CO 42-100201 to J-85-CO 42-100262	A6B	A3B	A-13	Flex	A6B
J-85-CO 42-100263 through J-170-CO	A6B	A3C	A-13	Flex	A6B
J-175-CO	A6B	A3D	A-13	Flex	Various
J-180-CO	A6B	A3D	A-13	Flex	Various
J-185-CO through J-210-CO	A15	A3D	A-13	Flex	Various

The variations in tail armament that began at Block 165 are interesting and little publicised. Beginning with serial 44-40521, Consolidated delivered 122 B-24J's with the light-weight M-6A 'tail stinger', a hand-held but hydraulically-assisted unit. These 122 aircraft were produced in small batches of not more than sixteen consecutive serial numbers, alternating with the regular A-6B unit. Also of interest is the fact that two B-24J's, serials 44-40936 and 44-41372, were delivered with the hand-held tail guns that were planned for wide use on the B-24L.

The J-CO aeroplanes deleted the port wing passing light during Block 50, added post-type fuel tank vents on the upper wing beginning with Block 75, deleted the familiar twin mast-type (D-1) pitots and static tubes in favour of a single flush-fitted G-2 type on the lower port side of the nose at Block 95, changed from twin wingtip lights to a single unit at Block 125, deleted paint at Block 145, went to hot-air de-icers at Block 150, deleted the tunnel gun scanning windows at Block 160, and adopted downward-opening nose wheel doors at Block 185 when the shift to the Emerson nose turret was made.

In September 1943, a separate parts plant at Fort Worth was completed and began supplying sub-assemblies to the main assembly plant, marking the beginning of Fort Worth's independence from Willow Run. The units from the new parts plant were assembled as B-24J-CF's while the steady supply of KD's from Ford continued to be assembled as B-24H-CF's. Thus H's and J's were produced alternately on the Forth Worth line from September 1943 to 2 May 1944 when H-CF serial 42-50451 was completed. Then, in a move destined to confuse the picture for years to come, the last fifty-seven Ford H knock-down kits were assembled as part of production block J-40-CF—complete with Emerson nose turrets, A-3D top turrets, natural metal finish, enclosed waist openings and no tunnel gun scan windows. (Sometimes these fifty-seven machines are referred to by a separate block designation, J-401-CF.) 'Normal' J-CF's adopted the Emerson at Block 45, deleted paint at Block 55 and never did enclose the waist or delete the scan windows. Adding to the confusion is the fact that these fifty-seven aircraft carried serials 42-50452/50508, which were a continuation of the last H-CF serials. Other J-CF changes included adding the post-type fuel vents atop the wing during Block 20, switching from mast-type to G-2 pitots at the start of Block 25, dropping the upper and lower wingtip lights in favour of a single unit in Block 30, and eliminating the

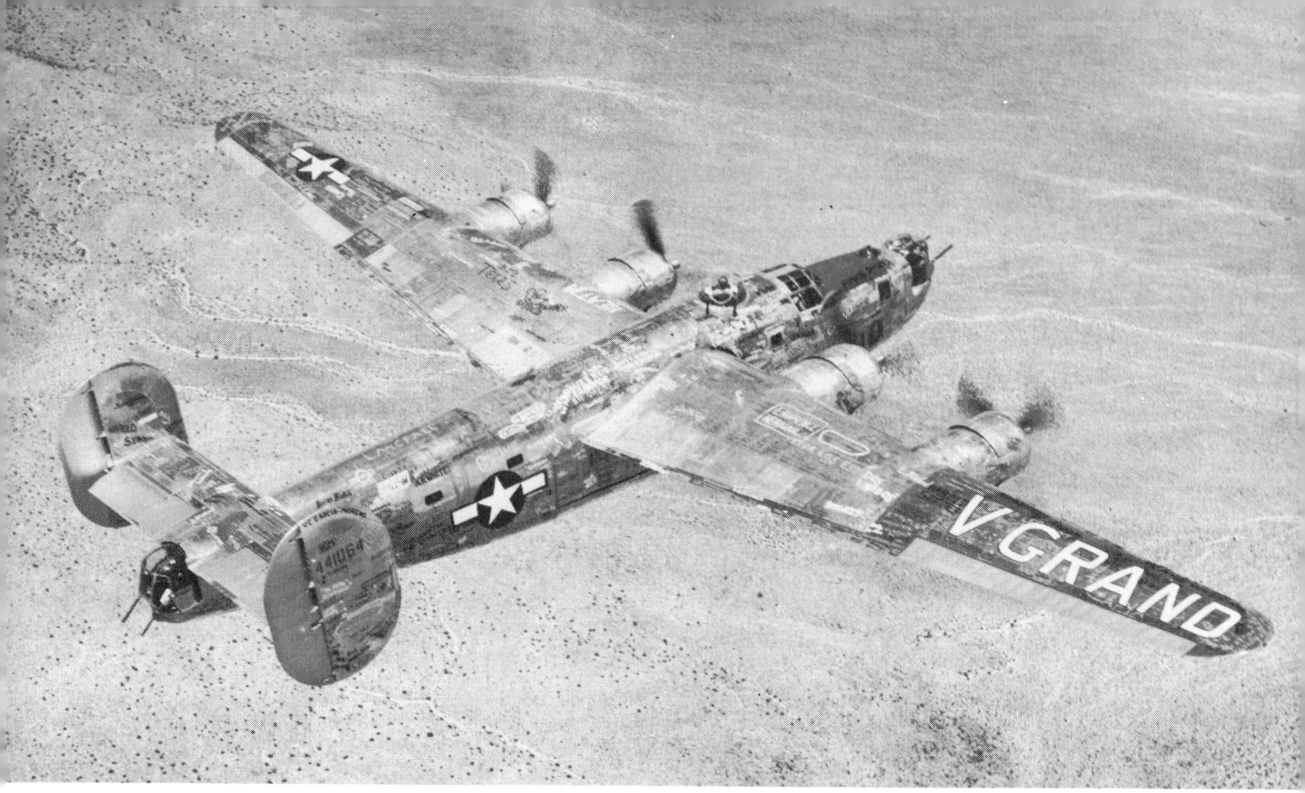

V GRAND
441064

28
BOISE BRONC

Above left: Autographed by dignitaries and workers alike, B-24JV GRAND, San Diego's 5000th B-24, served in the Mediterranean in all her informal glory. Twice the ship made emergency landings on Vis Island in the Adriatic Sea after losing engines on operational missions. / Consolidated

Left: B-24J-CO 44-40728 on Palawan, July 1945. This shows good detail of G-2 type pitot which replaced the familiar mast type at Block J-95-CO. Bombardier's windshield wiper also shows clearly in this photo. The bomb bay doors on this aircraft are not completely closed, which accounts for the gap between them and the fuselage. / USAF

Above: A B-24J-CF. Many in this series were Lend Leased to Britain and Canada. / Consolidated

Right: XB-24J 4Z-73130 with B-17 nose.

port wing passing light at the end of Block 55. The vestigial tunnel gun scanning windows continued to appear on the Fort Worth J's until production ceased.

Some misunderstanding has arisen because of published statements implying that individual B-24H's (and in some cases G's) were redesignated B-24J's in operational theatres when the C-1 autopilot and M-9 bombsite were installed in place of the previous A-5 and S-1 types. This was not the case; the C-1 and M-9 installation was a part of the changeover to J's on the production lines and, with the two exceptions noted below, there were no H-to-J, or G-to-J, redesignations after the aircraft left the factory. The first exception was B-24H-20-FO 42-95100, which was retained at Willow Run and became an XB-24J, the prototype for Ford's B-24J production. Still retaining its XB-24J designation, this plane was used by the General Electric Company after the war for experimental purposes and was eventually sent for salvage in March 1948. The other XB-24J resulted from B-24J-15-CO 42-73130 being assigned an 'X' designation when it was fitted with a B-17 nose in an attempt to improve the Liberator's forward-visibility problem. Later this same bird saw experimental service with the Air Force Special Weapons Command at Kirtland AFB.

Although somewhat out of context here, another case of altered identity involved the Liberators bearing serial numbers 41-29007 and 008. These aircraft were shipped from San Diego to the Douglas Tulsa plant as unassembled B-24D's, where Douglas used them as practice units to train production workers. Later accepted by the USAAF as B-24E's to fulfil contract requirements, the planes remained at the Douglas factory in the educational role until mid-1944. They were then brought up to B-24H-15 standard and turned back to the USAAF on 31 July 1944. A somewhat similar story involves a Ford machine, 42-7770, which was accepted as a B-24E but, as a company-retained aircraft, was eventually transformed into a B-24H and was then carried as such on USAAF records.

Still on the subject of confusion, a great deal of erroneous information has also appeared in print concerning the types of nose turrets used on the various batches of Liberators. Actually, the breakdown is fairly straightforward and we can now summarise, as follows, the distribution of this identification feature: all production B-24's equipped with a factory-installed nose turret used a version of the Emerson (cylinder shaped) unit except the following, which used the Consolidated (sloping front) type: B-24J-1-CO through B-24J-180-CO inclusive; B-24J-1-CF through B-24J-40-CF inclusive, less the fifty-seven B-24J-401-CF's bearing serials 42-50452/50508.

The B-24J's built by North American were generally similar to those of the other manufacturers, although small individual differences can be noted for recognition purposes. For example, all B-24J-NT's had the twin aileron tab feature, used boot-type de-icers, retained the twin wingtip lights and the port wing passing light, had flush type pitots and featured factory-installed bombardier scan windows in the nose.

Although such small differences between the Liberators produced by the various manufacturers may seem trivial, they were considered anything but trivial at the time. To the men who had to maintain the aircraft in the field, in fact, there were four different kinds of B-24's, each of which required its own manuals, parts lists and brand of profanity to fix. Ford models, including the Douglas and Fort Worth KD's, had interchangeable parts, as did the North American models, although these two types were not interchangeable with each other. The other two types were those manufactured at San Diego and Fort Worth, for while all Consolidated engineering was controlled at the home plant the lack of strict dimensional control kept these supposedly identical versions from being completely interchangeable. A tooling sub-committee of the B-24 Liaison Committee, formed in February of 1944, worked long and hard to establish true interchangeability among all manufacturers but had not accomplished its objective before production cutbacks mooted the point.

Top: The Fort Worth production line during September 1943, showing alternate assembly of J-5-CF and H-10-CF aircraft. Note use of Consolidated turret on the former and Emerson on latter. The Js in this photo are for Britain; aircraft in foreground is already lettered EV852. C-87 assembly line on left. /Consolidated

Above: The XB-24K at Eglin, April 1944. /USAF

Above: The Eighth Air Force experimental nose under study at San Diego. / General Dynamics

Below: XB-24K with modified nose section.

The XB-24K

Consolidated recognised at an early date that a single tail configuration for the Liberator might prove highly desirable, but because of the press of early production commitments it was not until 10 October 1942 that the first wind tunnel tests got under way on a model incorporating this feature. When they gave encouraging results, B-24D 42-40058 was selected as a test article. Known as the B-24ST (for Single Tail), this aircraft was fitted with the horizontal and vertical tail surfaces from a Douglas B-23 and first test-flown on 6 March 1943. The B-23 horizontal components were later discarded and replaced by equivalent C-54 units from the Douglas factory. The B-23 fin remained but was modified and a new rudder fitted. This version continued the flight test programme until June when, on the 29th of that month, the entire rear fuselage assembly of 42-40058 was removed and spliced to another similary-dissected Liberator, B-24-D-40-CO 42-40234. The latter, after its acceptance by the USAAF on 23 January 1943, had been retained by San Diego as a special 'proof ship' for MCR's[19] and, as a result, actually represented a much later version of the B-24 than was then coming off the assembly lines. (It had, for example, a Motor Products nose turret and R-1830-65 engines.) Now with its grafted-on single fin empennage, 42-40234 was given the designation XB-24K—although many San Diego personnel continued to refer to the 'B-24ST'. (Note: Official records retained the XB-24K designation although the aircraft itself was stencilled B-24K—no 'X'. Two other B-24's assigned to San Diego's Flight Test Department were involved in minor aspects of the single tail programme; these were 42-40175 and 42-73330.)

After a series of flights from San Diego, the first of which took place on 9 September 1943, the XB-24K was flown to the USAAF Proving Ground Command at Eglin Field for official tests. Eglin's experienced B-24 pilots—not usually given to superlatives in their reports—found it decidedly superior to any other B-24 they had flown. Handling characteristics and manoeuvrability were excellent, controls more sensitive, directional stability increased, and performance with two engines out on the same side was a great improvement over the standard B-24. Also noted were large increases in the fields of fire of the top turret, ball turret, tail turret and waist guns.[20] Without qualification, Eglin recommended on 26 April 1944 that 'an empennage of similar design be incorporated in all future production B-24 aircraft' and preliminary contract negotiations were initiated for an unprecedented order of 4,500 B-24K machines. In time this programme resulted in the B-24N. The sole XB-24K continued to be used as an experimental aeroplane and later, stripped to bare metal finish, received Bell power boost tail guns and a completely redesigned nose section based on the design submitted for USAAF consideration by the Eighth Air Force. In 1945 the ship was used to develop the 'flying tab' empennage controls later used on the B-36 and XB-46, and at some stage of the game acquired a cockpit canopy from a Douglas A-26. The 'Kay', an experimental aircraft in every sense of the word, was finally scrapped on 29 November 1945.

The B-24L and B-24M

By mid-1944 it was obvious that the two main production sources for the Liberator, San Diego and Willow Run, were capable of delivering more than enough B-24's to fulfil the needs of the USAAF and its Allies. Accordingly, B-24J assembly by Douglas at Tulsa, and North American at Dallas, was terminated. Fort Worth continued to build B-24J's until the end of the year—almost all of which were used to fulfil commitments to the UK—then it, too, withdrew from the production pool to concentrate on the B-32. But by mid-August, 1944, both Consolidated and Ford had shifted from the B-24J to the B-24L.

The main visible difference between the J and the L was in the armament fitted. With the L, Consolidated finally abandoned the open waist gun idea and went to an enclosed mount designated the K-5. At Block L-15-CO this was changed to the K-6 mount, which Ford had been using since the start of B-24H-20-FO. The variety of tail armament

Top: Number 5000 for Ford. The ship which was with 11th Bomb Group flew eight missions before war's end. / USAF

Above: Late B-24M-FO still sports tunnel gun scanning windows. / Via Hal Andrews

begun with the later J-CO's continued, however, and of the 417 L-CO's produced, forty-one had the M-6A stinger, 190 had the A-6B turret, and only 186 used the hand-held installation usually associated in print with the B-24L.

The reason for the variations in tail armament stemmed from a decision on 10 July 1944 to have San Diego, Willow Run and Dallas deliver aircraft to the modification centres without any tail armament at all. At the centres, small batches could then be processed to meet individual theatre requirements—e.g., Bell power boost guns for the Thirteenth Air Force, the A-6B turret for the Eighth Air Force, etc. Under this scheme, all aircraft delivered by San Diego without tail armament were to be designated B-24L, those delivered by Ford were to be designated B-24M, while deliveries by North American would be designated B-24N. This arrangement was actually put into effect for a brief period, and when it was cancelled shortly thereafter it left a rather untidy situation with regard to 115 Liberators which had been delivered to modification centres by Ford without tail armament and duly designated B-24M's. The confusion was resolved by a highly unusual Tech Order dated 10 October 1944, the text of which is reproduced here in its entirety:

> To provide proper classification and markings, the following B-24M airplanes will be redesignated as B-24L airplanes. All records, charts, stencilling, or painting on fuselages, and all other listings will be changed accordingly. AAF Form 110A and 110B will show the new designations when submitted.

AAF Serial Nos

44-49002 to -49073 incl.	44-49120
44-49075 to -49093 incl.	44-49121
44-49095	44-49123 and -49124
44-49097 and -49098	44-49129
44-49100 to -49102 incl.	44-49132 and -49133
44-49104 and -49105	44-49152
44-49109 and -49110	44-49154
44-49112 and -49113	44-49161 and -49162
44-49115	

Other exterior changes during the L and M production runs included:

L-5-CO—addition of aileron tab on port wing and factory-installed bombardier scan windows.
M-15-CO—addition of static discharge wicks on wingtips.
M-20-CO—provision for interchanging navigator's astrodome with a flat plate glass unit.
M-25-CO—ditching provision for crew provided over the aft bomb bay.
L-1-FO—astrodome/flat plate interchange feature.
L-15-FO—increased span of aileron tabs (FO machines, as previously noted, had had two tabs since the H series), aft bomb bay ditching provision, wingtip static discharge wicks.
M-20-FO—complete pilot's greenhouse rework, similar to type used on the B-24N.

Interior changes included installation of a flux-gate compass starting at L-5-CO, an all electric bomb release system at M-5-FO and M-15-CO, and non-metallic, self-sealing, 42-gallon oil tanks at M-15-FO and M-20-CO.

The B-24N

During World War II, complaints from the operational theatres concerning lack of aircraft performance could often be traced to the fact that the aircraft was being operated under conditions far in excess of design specifications. Yet when changes were made to improve lagging performance, the theatres invariably used these improvements to further increase maximum loads instead of taking the improvement in terms of the increased performance that was originally requested.

Under this axiom, the B-24J was as a routine being flown on missions at gross weights in excess of 36 tons and flying characteristics suffered accordingly. Liberator controls had always been heavy, and continual addition of weight made them worse. It was extremely difficult to fly a tight formation with the B-24J, and so tiring that many pilots found it physically impossible after a few hours. The slow rate of roll, an inborn characteristic of the

Top: The 7th-from-last Ford B-24M, 44-51922, at Wright Field on 27 June 1947. Note V windshield, buzz number and painted inboard surfaces of nacelles. Guns have been removed; — 75 engines installed. / Via Hal Andrews

Above B-24M-15-CO Tanker being used to haul gas into Kunming, China. Photo taken at Myitkyina, Burma in late 1944. Many B-24's were used as tanker aircraft without full-scale conversion to C-109 standard. / Ken Sumney

Liberator because of the large wing span, had been made slower by the addition of the outer wing tanks. Initial climb of a combat-loaded B-24 was slow, usually taking about six minutes to reach the 1,000 ft mark. Visibility from the flight deck was inadequate for formation flying. Because of the manner in which the side windows sloped inward, the pilot and co-pilot had to be seated low in order to have enough head room. This made it difficult to see over the instrument panel. In addition, the astrodome and the top of the nose turret were directly in front of the windshield which, to begin with, was considered too narrow and already partly blocked by the above-the-dash location of the compass. Furthermore, the side windows were so small that it was dangerous for the pilot to put his head through the window while taxiing. Lack of visibility from the nose compartment was equally serious—the only way the bombardier could adequately see was to get down on his hands and knees, and it was impossible for the navigator to help in target identification because of the lack of window area.

The Eighth Air Force experimented with various means of reducing the weight and visibility problems of the Liberator, including the removal of all ball turrets in July 1944 and the addition of various additional window areas in the nose. General Doolittle eventually rejected the Liberator on the grounds of too much weight and too little visibility and it was his intention, if sufficient B-17's could be obtained, to convert the entire Eighth to the Fortress.

The Liberator's handling and visibility problems were due in large measure to the nose turret, which was heavy and produced excessive drag. This problem had already spawned a variety of makeshift solutions at home as well as overseas. The Eighth equipped two B-24's with Bell power-boost turrets and sent one back to the United States for evaluation, where it was promptly rejected. The Materiel Command tried several approaches including the experimental installation of a B-17 nose on the B-24 mentioned earlier. But none of these solutions was completely satisfactory.

Some of the individual problems were treated piece-meal in the B-24L and B-24M including, as previously mentioned, additional transparent panels in the nose, lighter weight tail armament, a complete rework of the windshield and flight deck canopy, more powerful engines, trim tabs of increased area on both ailerons and revised aileron control linkage to lessen the required stick forces. But it was recognised that the final solution lay in a fairly extensive redesign of the Liberator which would reach series production as the B-24N. Although only eight examples of this version were produced, it did represent the ultimate wartime development of the B-24 by the USAAF and it is described here, not as what might have been, but what almost was.

In outward appearance the most obvious changes were the single fin empennage, the Emerson ball nose turret, and the remotely operated and sighted, ball mounted, tail guns. (This tail gun installation, which appeared on the prototype, was a production expedient to be replaced at the factory and in the field as soon as volume production of Southern Aircraft Model 7 light-weight turrets permitted). The Davis wing was unchanged. Horizontal stabiliser area and span were increased, vertical stabiliser area decreased. The B-24N, measured from ground to rudder tip, stood 26 ft 9 in., or 8 ft 10 in. higher than the B-24M. Fuselage height and width were unchanged, but over-all length was reduced to 67 ft 1 ½ in. in comparison with 67 ft 7 5/8 in. for a B-24J equipped with a Consolidated nose turret. Engine nacelle over-all length was increased 5 ¼ in. ahead of the wing in order to mount Pratt & Whitney R-1830-75 engines with 42-gallon self-sealing, non-metallic oil tanks.

The power plant installation included nose-mounted Stromberg SF 17 LN-8 magnetos, Stromberg PD 12 F-8 carburettors, and R-1 high speed 300 A generators. Engines in the first production B-24N's were to be equipped with a cast ignition harness which was considered a production expedient to be replaced, both at the factory and in the field, with titeflex ignition harnesses. A turbo bypass door in the engine air duct system

Above: The XB-24N. /USAF

Left: As noted in the text, the bombardier's view forward was somewhat restricted. /USAF

permitted take-off without use of the B-31 type turbo-supercharger, allowing cooler cylinder head temperatures and reducing the possibility of cylinder detonation on take-off. This gave a noticeable increase in engine torque because the turbo-supercharger and its accessories were inoperative and not taking their normal percentage of engine power output. The built-in characteristics of the R-1830-75 engine gave an increase of 100-150 hp per engine at take-off manifold pressures and war-emergency power. At cruising manifold pressures the -75 remained the same as earlier engines in rated horsepower.

The Liberator had had a relatively minor but persistent engine heating problem throughout its career, complicated by the interaction of airflow over open cowl flaps with the twin vertical tail surfaces. Early aircraft were equipped with 9½ in. flaps which opened to a maximum of 30°. Subsequently, flaps were increased in size to 11 in. and the full-open position reduced to 22°. The next version consisted of 10 in. flaps, with the top four opening to only 12½° while the bottom four were opening to 22°. The B-24N variation consisted of six adjustable 10 in. flaps, two fixed 10 in. and two 'false' flaps which formed part of the intercooler duct and air intake assemblies. The two top and four lower adjustable flaps opened to 22° while the two fixed flaps were set at 2½°.

Larger capacity Stewart Warner 900C heat exchangers on the N made available considerably more heat for wing and empennage anti-icing and cabin heating, making possible the elimination of the troublesome spotheaters of earlier Liberators. The windshield was double thickness, with ducts for circulating heated air between the glass for defrosting. Pilot and co-pilot inner windshields could be replaced with 40 lb bullet-proof glass panels. Wing and empennage de-icing was by circulated heated air, with the former circulating through the wing and out at the aileron hinges for hinge and aileron de-icing.

The interior arrangement placed the navigator facing forward in the nose compartment with an alternate position behind the pilot on the flight deck. The radio operator's position was revised to lessen interference with the top turret, and ditching positions for the crew were moved to the command deck above the aft bomb bay. The cockpit canopy was redesigned to provide greater visibility and a 'push-out' capability to provide overhead escape routes for both pilot and co-pilot. Instrument panel and centre pedestal were redesigned for greater efficiency and pilot and co-pilot were provided with C-1 automatic pilot and formation stick controls. The flight deck compartment was strengthened structurally to give better protection on forced landings. This included additional reinforcement of the fuselage structure forward of the top turret, which had shown a tendency to fail during crash landings of earlier Liberators, allowing the turret to move forward with lethal consequences. Ammunition capacity for waist guns, tail and nose turrets was increased slightly, while bomb load remained the same as in previous B-24's.

It is interesting to note that the single tail assembly of the B-24N was designed so that the horizontal stabiliser attached to the fuselage with the same fittings used to secure the original type, while the vertical fin was, from a structural stand-point, independently attached to the new stabiliser. Thus it would have been entirely feasible to undertake a programme of converting twin-tailed Liberators to the more desirable single-tail configuration without major structural problems. The possibility of just such a retrofit programme was not lost on CVAC's engineers, and by the early spring of 1945 the enterprising Technical Operations Section of the Eighth Air Force had started the conversion of a B-24M. (It remained uncompleted at the end of the war.)

Since the XB-24N was equipped with servo tabs on all controls, great improvement was evident in this area and it was felt that pilot fatigue problem during formation flying had been licked. Take-off performance was impressive, with the aircraft using an average of 400 ft less than a B-24G used for comparison purposes. Climb was greatly improved, and all altitude figures were considerably better than production Liberators.

Right: Prototype B-24N nose package installed on B-24G 42-78399. / Emerson Electric

Below: As part of the campaign to reduce the weight of the Liberator, the old tunnel gun idea was resurrected and in October of 1944 a B-24L-10-CO, 44-41627, was fitted out with ring-mounted twin fifties in Bell E-11 adapters as shown above. The installation also brought back scanning windows for the gunner. Although the set-up saved a welcome 775 pounds in weight, sighting was '. . .so poor that even the tow targets used for testing had to be pointed out to the gunner. . . it was virtually impossible for him to pick up a target when in firing position.' Compare with the regular Briggs A-13A ball turret, shown in extended position in companion photo. / USAF

Left: The second Eighth AF attempt to incorporate the Bell power boost unit into the Liberator nose. This version was on an ex-44th Group aircraft, *Shoo Shoo Baby* (B-24H 41-29208]. / General Dynamics

Below right: B-24N tailgun. / General Dynamics

448753
449726

However, although noted as being better than in previous B-24's, the pilot's forward visibility in a climb was still rather poor and flight characteristics with an outboard engine windmilling were still considered 'undesirable and dangerous'. The new-fangled cockpit heating system didn't work at all above 5,000 ft on the test aircraft and, as the test pilot felt he should remind the readers of his report, 'It is very cold at 23,000 ft.' Cockpit noise and vibration levels were objectionable at high power settings, as always, and the cowl flaps still tended to creep open.

The XB-24N arrived at Wright Field on 20 November 1944 to begin its test programme, which got under way on 16 December. During all but one test flight the ship was equipped with a false barbette located on each side of the fuselage, below and in front of sighting blisters installed in the waist windows, to simulate an armament change under consideration. These were then removed for one flight and the regular waist guns and ball turret were installed to determine the difference in drag of the two installations. It was found that the latter configuration, with ball turret retracted, was 3 mph faster than the barbette version but up to 6 mph slower than the barbette version when the ball turret was extended. On balance, it was felt that the barbette installation was a good compromise.

The testing programme was delayed by bad weather and the discovery of an airspeed calibration error which necessitated repeating all the speed and power runs. No explanation could be found for the fact that it was impossible to keep the XB-24N in straight flight below 135 mph with one engine windmilling, while the XB-24K had shown no such problem. The mystery sent the aircraft to San Diego, where a fin flare (which Ford had omitted) was added and the rudder from the second RY-3 was installed to check for possible warping of the original. Neither of these changes cured the problem, however, and San Diego continued to search for the answer while Ford planning went forward for full production. The first YB-24N (44-52053) was accepted on 30 May 1945 and the remaining six the following month. Thousands more were cancelled. Who knows—perhaps if they had all been built the problem of how to keep Liberator crews warm would have been solved somewhere along the line!

The XB-24P and XB-24Q

After its delivery on 15 February 1943, B-24D-CO 42-40344 was retained by Consolidated for test purposes until July. It was then made available to the Sperry Gyroscope Company for fire control research and received the designation XB-24P. After the completion of its Sperry assignment, Wright Field retained custody of the aeroplane and used it for miscellaneous assignments until December 1945, when it was flown out to pasture at Kingman, Arizona, still carrying its designation as the one and only XB-24P.

The XB-24Q began life on 8 November 1944 as B-24L-FO 44-49916. Also a test aircraft for nearly all of its career, the ship was first loaned to the Emerson Electric Company at St Louis. It was not until 30 July 1946 that it became the XB-24Q while in the hands of the General Electric Company at Schenectady, New York, whose personnel modified it to test the radar-controlled tail gun installation that the company was preparing for the B-47. At one time the aircraft also had an I-40 jet engine installed in the tail. The end of the XB-24Q is somewhat of a mystery, for it is recorded as being scrapped at an unidentified overseas location as of 16 August 1948.

Navy Liberators

The PB4Y-1

Almost immediately after the US entered the war, German U-Boats moved into close American waters with devastating success. Caught badly unprepared, without proper plans or equipment to meet the onslaught, US reaction was further weakened by an interservice rivalry over which Service was charged with the responsibility for the antisubmarine air effort. The Army Air Corps was given the job initially, but there were too many other claims on the few suitable aircraft that were available, and the first squadrons assembled to meet the German challenge were indeed a rag-

Right: Test installation of fan-assisted cooling. / Via Joe Famme

Below: B-24N-CO flight deck. / General Dynamics

Above: Close up of first B-24 barbette turret. / Bendix Corp

Below: YB-24N-FO 44-52056. / Consolidated

Above: Close-up of the XB-24Q tail area. / Via Hal Andrews

Below: The XB-24Q showing waist sighting blister. / Via Bob Esposito

tag lot. By the autumn of 1942, however, the first few B-24D's assigned to antisub duty had demonstrated that the Liberator was the best US aircraft for the job—a fact the British had recognised considerably earlier—and from that time on the B-24 became the foundation of the US antisubmarine effort. In the autumn of 1943 the USAAF relinquished its antisub duties and turned over some forty-five sub-hunting Liberators to the USN. These joined the 167 Liberators already in Navy service at that time (the first having been delivered in September 1942) which also were being used for antisub work as well as other Navy duties. All USN Liberators were designated PB4Y-1's, regardless of their USAAF series designation.

Early-series PB4Y-1's included fifty-one accepted in 1942 and 264 in 1943, plus the forty-five taken over from the USAAF as the antisub job changed hands. A number of these ex-B-24D aircraft were modified to take an Erco (Engineering and Research Co.) ball turret in the nose. Deliveries of later variants brought the PB4Y-1 total to 977 by January 1945 when the Navy's last twin-tailed Liberator, a B-24M, was handed over to that service. It should be noted that *all* PB4Y-1's were San Diego-built aircraft except one, the exception being a single test B-24G (42-78271) which was assigned BuNo 38782.

A number of PB4Y-1's were converted to a crude 'assault drone' configuration in late 1943 and 1944. Loaded with high explosives, these aircraft were taken aloft by a pilot and co-pilot who climbed to altitude, set up the required power conditions, passed control of the Liberator to a control plane and then bailed out. Created in the European theatre originally for use against the concrete sub pens on the French channel coast and later against V-1 and V-2 launching sites, these 'flying bombs' were not particularly effective and the USN operational programme was discontinued after two operational sorties. (The first one, BuNo 32271, blew up and the second missed its target. Similar USAAF efforts with B-17s were just about as successful.) The ideas pioneered in 1943, however, would eventually lead to a much more sophisticated, completely remote-controlled Liberator type nearly a decade later. This development is covered in the PB4Y-2 section below.

The PB4Y-2

Although it did not carry the Liberator name, the Privateer was nonetheless a full-blooded member of the Liberator family, having stemmed from three XPB4Y-2 prototypes converted from PB4Y-1's during 1943. These aircraft, BuNos 32095, 096 and 086, were ordered to the new configuration on 3 May. Work was performed by Consolidated and the three prototypes flew on 20 September, 30 October and 15 December, respectively. Although it had been a feature of the original PB4Y-2 design, the three prototypes were delivered and first flown *without* the single tail that characterised the production version of the Privateer. In late 1943 all three XPB4Y-2's were returned to San Diego for the addition of this change and the aircraft were re-delivered to the USN on 2 February 1944. The first two prototypes retained their standard B-24 power package, while the third had the engines and nacelles planned for production.

The mission initially considered for the PB4Y-2 was for long-range airborne radar search. Referred to unofficially as the 'Two by Four', the Privateer (CAC Model 40, later changed to CVAC Model 100) differed from the standard Liberator in having a 7 ft fuselage stretch forward of the wing leading edge, resulting in an overall length of 74 ft 7 in. (Or 74 ft 8¼ in. when equipped with a Motor Products A-6 type nose turret, as some of the earlier PB4Y-2's were.) Although the basic Liberator wing was unchanged, the unsupercharged R-1830-94 engines were housed in revised ovoid nacelles with their long axis vertical instead of horizontal. The outer wing fuel tanks were not installed, since the use of four bomb bay cells provided sufficient total capacity (4,108 gallons) for the longest mission the aircraft was scheduled to perform. Completing the exterior changes, a single fin empennage of rather startling size was added, increasing the overall height to 29 ft 1 5/8 in.[21] The span of the horizontal stabiliser was 34 ft 6 in. Privateer armament included twin Martin

Top: **A Navy Liberator in flight.** / General Dynamics

Above: **A USAAF anti-submarine B-24D,** ***Night Mission*****, serial 42-40891.** / Via Hal Andrews

Top: This is first PB4Y-1 to be received by the USN and was used for Inspection Trials. Initial PB4Y-1's carried both Army serials and Navy BuNo's. / USN

Above: An early PB4Y-1 showing Erco nose turret modification on B-24. / Via Hal Andrews

Top: The second XPB4Y-2,. This X model retained the original B-24 power package, as shown here. Faded name on fuselage reads *Patuxent River Wart Hog*. /USAF

Above: The first XPB4Y-2, BuNo 32095, shown here during the brief period during which the prototypes retained their twin tails. /General Dynamics

top turrets, a Consolidated tail turret, and an Erco 260TH Tear Drop turret (two ·50 calibre guns) at each waist position. (It is of interest to note that the adding of these Navy turrets completed a full cycle for the Liberator, for the original Model 32 mock-up release schedule of 23 January 1939 called for XPB2Y-1 turrets to be used at the waist positions.) Nose armament on the XPB4Y-2's was a standard Liberator CAC nose turret, while most production aircraft were fitted with the Erco 250-SH-3 ball unit. On a limited number of PB4Y-2's this turret was installed with a 13° forward cant to increase the field of downward fire. Extensive search radar equipment was carried, the antennae of which accounted for the variety of bumps and protrusions that decorated the Two by Four.

Navy orders for the Privateer included 660 in 1943 (BuNos 59350/60009) and batches of 150 (BuNos 66245/66394), 260 (BuNos 66795/67054), and 300 (BuNos 76839/77138) in 1944. Of the 1943 order, BuNo 59554 was destroyed prior to Navy acceptance and no replacement was ordered, so that 659 was the total delivered. The remainder of the 1944 orders was cancelled in 1945 after eighty aircraft (BuNos 66245/66324) had been delivered. Thus the total number of PB4Y-2's received by the USN, excluding the three prototypes, was 739. All were built at San Diego and given post-production modification at Goodyear Aircraft's Arizona facility near Phoenix.

One hundred and twelve examples of a cargo version of the Privateer, which received the designation RY-3 (C-87C or Convair Model 101), were ordered in March 1944. Also built at San Diego, four were delivered in 1944 and twenty-nine more in 1945 before the balance of the order was cancelled. RY-3 BuNos were 90020/021, 023/050 and 057/059. (Cancelled numbers were 90051/056 and 060/131. One Ry-3, ex-BuNo 90022, was retained by Convair.) Of the thirty-three delivered, twenty-six went to the UK as the Liberator C.IX where they were assigned serials JT973, JT975/998, and JV936.

Although the last Privateer was delivered in October 1945, the type was active in Navy service as the P4Y-2 until well into the 1950s, serving in a variety of roles which included hurricane hunting and photo reconnaissance (as the P4Y-2P) as well as long-range patrol. One Privateer from VP-26 was lost on 8 April 1950 after being attacked by Russian aircraft over the Baltic Sea. The aircraft also saw service in the Korean War where, on 12 June 1951, VP-772 Privateers flew the first of many flare-dropping missions to provide illumination for Marine ground attack aircraft.

In 1952 the Naval Air Development Center converted a number of Privateers to drones. Designated P4Y-2K's, these aircraft featured a camera pod on each wingtip to photograph missiles fired at the drone. Assuming the missile missed, the miss distance and other flight data could be calculated from the pictures. P4Y-2K's could be flown up to a range of thirty miles from the controller's location.

The USAAF quickly phased the B-24 out of service after V-J Day, with its numbers dropping from 4,986 as of June 1945 to 3,458 on 30 September and 1,103 on 31 December. By 31 December 1948, only two examples were listed in the USAAF inventory. On the other hand, USN use of the PB4Y-2 Privateer was just coming into its own by mid-1945 and these aircraft extended the era of the aeroplane's first-line service for another eight years, or well into the jet age. As late as 1961, an all-red QP4Y-2 was being used to ferry personnel between the Navy missile installation at Point Mugu, California and St Nicholas Island. During the same year Burmese Air Force fighters shot down a Privateer being used by Nationalist Chinese forces to drop supplies to guerrillas in the Shan States region. Other Privateers have remained active in the service of some of the Latin and South American nations.

Foreign Allocations

British interest in the Liberator was firmly established by the spring of 1940 and before the end of that year the British Purchasing Commission was seeking permission to negotiate, in addition to the somewhat smaller orders already placed, contracts that would run

Top: The single fin configuration of the PB4Y-2 greatly improved the field of fire for the Erco waist turrets. /USAF

Below: Not all production PB4Y-2's used the Erco ball nose turret. /Emerson Electric

The Superchief
740

Above left: PB4Y-2 Privateer on Palawan showing area of fuselage stretch and antennae details. /USAF

Left: BuNo 66324, the last PB4Y-2. Note the difference between the props on this low altitude Privateer and the 'paddle' props installed on Liberators. / General Dynamics

Top: The first RY-3, BuNo 90020. / Consolidated

Above: RY-3 BuNo 90026, alias JT978. / USN

British Liberator deliveries into the thousands. While events soon overtook these plans, the groundwork had been laid to establish the UK as a major user of the aircraft. This policy reached fruition under the Lend Lease programme.

Under Lend Lease, Britain received 1,865 Liberator bombers and twenty-four Liberator transports, Canada eighty-eight B-24's,[22] and Australia 275 B-24's for a total of 2,252 aircraft. In addition, the RAF and RAAF received a total of fifty-five B-24's as operational theatre transfers. The RAF also acquired twenty-six RY-3 transports. Adding these to the 112 LB-30A's, LIB I's and LIB II's that Britain acquired as a result of pre-Lend Lease agreements gives a grand total of 2,445 Liberator-type aircraft in UK service. Distribution by series is shown in the following table:

	RAF	*RAAF*	*RCAF*	*Total*
LB-30A	6	0	0	6
LIB I	19	0	0	19
LIB II	87	0	0	87
B-24D	366	12	19	397
B-24G	8	0	0	8
B-24H	22	0	0	22
B-24J	1157	145	49	1351
B-24L	355	83	16	454
B-24M	0	47	4	51
C-87	24	0	0	24
RY-3	26	0	0	26
	2070	287	88	2445

These RCAF figures do not reflect the sixty RAF Liberators (all B-24J) which were held in Canada under the British Commonwealth Air Training Plan.

In the early days of short supply, Lend Lease commitments were met on a catch-as-catch-can basis and deliveries seldom resulted in a run of more than a few consecutive USAAF serial numbers. Later on, longer runs appear and it seems certain that an attempt was made to achieve a measure of standardisation in the series, as well as the manufacturing source, of the Liberators provided. The RAAF, for example, received San Diego and Dallas J, L and M machines exclusively while shipments to the RAF were made up of a large percentage of B-24J's built at Fort Worth. As examples of the latter, the 452 B-24J-CF's between 44-44049 and 44-44500 were all sent to Britain or Canada, and 362 of the 500 B-24J-CF's 44-10253/44-10752 were also divided between those two countries. Only one B-24H (42-64432) was supplied under Lend Lease, although the RAF subsequently picked up an additional twenty-one of this version via theatre transfers. The eight B-24G's were acquired in like fashion.

BRITISH LIBERATOR SERIAL ALLOCATIONS (Includes numbers not actually assigned to aircraft)		
AL503-667	FP685	LV336-346
AM258-263	JT973-999	TS519-539
AM910-929	JV936-999	TT336-343
BZ711-999	KE266-285	TW758-769
EV812-EW322	KG821-KH420	VB852
EW611-634	KK221-378	VB904
FK214-245	KL348-689	VD245
FL906-995	KN702-836	VD249
	KP125-196	

Until the last three months of 1943, all Lend Lease deliveries went directly to the UK except for eight aircraft that were handed over in North Africa. From October 1943 on, however, the majority of deliveries were made to locations outside the UK itself.

Equivalent nomenclature was as follows: YB-24 = LB-30A; LB-30B = LIB I; LB-30 = LIB II; B-24D = LIB III; B-24H = LIB IV; B-24D with Coastal Command modifications = LIB V; B-24J (plus some G and H transfers) = LIB VI; C-87 = LIB VII; B-24L (plus some late J) = LIB VIII; and RY-3 = LIB IX.

The following Squadrons used the Liberator as primary equipment during the war period:

RAF—Nos. 8, 37, 40, 53, 59, 70, 86, 99, 104, 120, 148, 159, 160, 178, 200, 203, 206, 215, 220, 223, 224, 232, 243, 246, 301, 311, 321, 354, 355, 356, 357, 358, 547 and 614.

RAAF—Nos. 12, 21, 23, 24, 25, 99 and 102.

Only two B-24's are known to have been officially transferred to French control during the war. On 1 April 1944, an ATC crew delivered B-24D 41-23913 to French forces in Dakar and on 9 April 1944 B-24D 42-40171

Top: The RY-3 (BuNo 90043) instrument panel. (General Dynamics)

Above: A P4Y-2K drone, BuNo 59759, in 1961.

Top: KH258 at Aktab, India in 1943. /Ken Sumney

Above: A B-24J-80-CF in the RAF. /Via Hal Andrews

Top: RAAF serial A72-116, ex-AAF B-24L-10-CO 44-41631. /RAAF

Above: A formation of 3 RAAF Liberators, ex-AAF serials 44-40651, 40871 and 41403. /RAAF

was transferred to French control at an unknown North African location.

The Russians got a single Liberator—by default. B-24D 41-11820, assigned to the Air Transport Command, was forced down in Siberia while carrying a load of VIP's. Since repair and recovery by US forces was next to impossible, the aircraft was placed on the Lend Lease accounts at the then-standard price of $394,084.90. This brought a howl from the Russians who did want the aircraft but didn't like being charged full price for what they pointed out was obviously used merchandise. The US officials readily agreed, and reduced the price to $340,346.04. Which, of course, was never paid either.

Four B-24J's were transferred to Yugoslavia with considerable fanfare in February 1944. These aircraft, flown by Yugoslav crews, operated with the 376th Bomb Group. Two more Liberators may have been similarly diverted in April 1944, but this has not been confirmed.

The first six B-24's to be assigned to the Chinese Air Force came out of the Tenth Air Force theatre reserve in January 1944. In April of the same year four more were transferred. These ten aircraft were used in an attempt to train Chinese crews to fly heavy bombardment aircraft. The programme did not pan out, and all ten Liberators were returned to Tenth Air Force control the following August. In May 1945 China received three B-24's outright, and was given thirty-four more in July. It is believed that these thirty-seven aircraft were all B-24M's.

Four Liberators landed in Turkey after the HALPRO raid on Ploesti during the night of 11-12 June 1942. (These were *Brooklyn Rambler,* 41-11596; *Blue Goose,* 41-11597; *Little Eva,* 41-11609; and *Edna Elizabeth,* 41-11622.) Some months later, *Brooklyn Rambler* rambled again when its pilot, 1st Lt Nathan J. Brown, took off from a field near Ankara in a daring escape from under the nose of the Turks. The Turkish government indicated more interest in getting the plane back than the men, but the ship had suffered nose damage during landing on Cyprus which was beyond the Turkish capability to repair. In the highly unusual settlement which was arrived at, the men rejoined the US forces and, on 22 February 1943, the Turks were permitted to fly the aircraft to Gura Air Depot in Eritria. There, on a cash-and-carry basis, Douglas Aircraft employees spent 8,156·5 man-hours repairing the ship before it was returned to its Turkish crew on 27 March. The total bill was $16,182.50, plus parts. It is not known how many other force-landed Liberators, if any, the Turks may have ultilised for their own purposes. Seven, for example, landed on Turkish soil after the low level Ploesti mission of 1 August 1943. (Including 41-23754, 42-40267, 544, 608, 744 and 777.)

Portugal acquired five Liberators in similar fashion, utilising four B-24D's (including 41-23740 and 42-40133) and one PB4Y-1 which made forced landings within its boundaries. All of these were used for transport service.

The South African Air Force had two operational squadrons equipped with RAF-supplied Liberators. These squadrons, Nos 31 and 34, were formed in the spring of 1944 and began operations from Foggia, Italy, in June of that year as part of No. 205 Group. The SAAF B-24's suffered heavily during the effort to drop supplies to the forces which inaugurated the premature Warsaw uprising in August 1944.

Some Notes on Production

The first Liberators, like other aircraft of the 1940-1 period, were mostly hand-built on an individual basis. As production orders grew, so did the amount of production tooling that could be economically justified and employed. By 1944 true mass production had been achieved; nearly 600,000 separate pieces of material were flown into each Liberator in a river of parts that began in the shops of the smallest sub-contractors and terminated at the end of the assembly lines of Consolidated, Ford, Douglas and North American. The problems to be overcome in maintaining this continuous flow were many, but one of the most important was 'production engineering'—i.e., breaking the aircraft down into small enough sections so that sufficient

Top: Liberator GRV, serial FL927. / Via Hal Andrews

Above: An early (FL series) LIB III. New Delhi, India, 1943. / Ken Sumney

Above left: FL910 in flight. An early Liberator III, this aircraft has neither Bendix remotely-sighted turret nor Sperry ball. However, note aft entrance hatch has been raised so that the tunnel gun can be swung into firing position.

Left: Flight over. FL910 was damaged by blast after attacking a U-Boat in 1942. It was brought back over 800 miles of sea during which time it persisted in climbing despite the efforts of both pilots. As a landing was being attempted the control lines snapped. Out of control, the aircraft struck a landing light and disintegrated. This is one of the pieces. /USAF

Above: KH228 with radome extended. /USAF

Below: The 'San Diego Hull and Wing Department' on 25 March 1940. /USAF

manpower could work on construction at the same time. Eventually, for example, each Consolidated B-24 nose section began as ten separate panels, each on its own assembly line. These ten lines merged into five, panels merged into subsections and finally everything came together as a single nose section complete in every detail. The same panel-subsection-section approach was taken throughout the aircraft so that the 3,000 ft San Diego final assembly line contained only thirty-one work stations. This assembly line, which held fifty-four Liberators, each canted at a 45° angle, moved at a constant speed (7 in. a minute) along with the work platforms which gave workers access to all sides of the ship. Scheduling was such that production control knew a wing centre section laid down in the jigs at 12 noon on 1 August would be ready for fuselage mating on 9 August at 2.43 am and would be rolled out as a finished Liberator at 4.53 pm on 14 August. Such mechanisation helped reduce the number of man-hours needed to build a Liberator from 40,000 in 1943 to less than 8,000 in 1945.[23]

The most dynamic B-24 production story, however, involved the Ford Motor Company's determined campaign to prove that true mass production of aircraft could be accomplished using the techniques of the car industry.[24]

The Willow Run story began even before Ford signed its original contract as a parts supplier. During March and April 1941, a team of 200 Ford representatives spent weeks at San Diego. They photographed every drawing of every different part and assembly going into the bomber. They made copies of all blueprints, bills of materials and engineering releases and made reproductions of the loft boards used by Consolidated. When they returned to Detroit, it took two freight cars to transport the material they had accumulated. They believed they had enough information to get Liberator production under way. No sooner had they returned to Detroit than their troubles began.

Much of the difficulty lay in the B-24 itself—the bomber simply was not ready for mass production in the sense that Detroit used that phrase. In addition, the Consolidated drawings turned out to be difficult to work with. Whereas Consolidated used fractional dimensions, Ford used decimal notations entirely. Moreover, the Consolidated draughtsmen all presumed that their drawings would be interpreted by experienced foremen and made use of all sorts of signs and symbols without amplification. Worse yet, the Ford engineers found numerous discrepancies between the duplicate loft boards and the detailed drawings of parts they had brought from San Diego. The Consolidated engineers knew of these errors in many cases but, under tremendous pressure, they customarily left it to the skill of their production men to reconcile them on the line.

For the engineers in Detroit, who were trying to prepare precision tools for mass production, such discrepancies were devasting. They decided they would have to re-do every one of the 30,000 drawings.

Sometimes the engineers working on a particular part would discover no drawing for it among the reams of paper brought from San Diego to Detroit. On one occasion, when a Ford draughtsman was unable to locate an adequate drawing for the B-24 toilet paper container, he wired CAC requesting further details. To his surprise the reply came back that there *were* no drawings. Consolidated had found it cheaper to buy the item in a local ten-cent store.

For nearly a year after the drawings were first brought back, a tool design group with over 1,000 men worked continually to prepare the jigs, fixtures and dies with which Ford hoped to achieve wonders of production. The management knew there were risks in planning tools while the aircraft itself was still being engineered for production—any subsequent design changes would force them to scrap production tools already made. But the risks seemed worth taking in the interest of speed, and by April 1942 one whole set of production tools was ready for use.

As it turned out, the decision was a wise one even though 15% of the tools had to be scrapped or reworked because of design changes. It was only because Ford had fabricated a large number of tools well before

Above: B-24-D-5-CF's in production. This camouflage scheme was standard factory finish for USAAF antisub Liberators during this period. / Consolidated

Right: Ford assembly jigs for mating fuselage waist and tail sections. / Ford

the assembly line was scheduled to start that the production men had time to discover some serious shortcomings that had to be worked out before production could begin. In fact, it gave them time to learn how faulty their basic assumption had been. Ford's whole programme rested on the premise that the production techniques of the automobile industry could be applied directly to aircraft. Experience showed this was not quite true.

The Ford engineers had planned to use dies far more extensively than was customary among the old-line aircraft firms. Once set up, tested, gauged and put into action, high-speed presses manned by relatively unskilled employees could turn out extremely accurate parts in large quantities and in very little time.

In practice, the use of dies proved disappointing. To begin with, the production men had to learn from bitter experience that the dies scratched and defaced the surface of aluminium, which was considerably softer than steel. A trial of chrome-plated dies proved abortive. After much experimentation, it was found that highly polished steel dies would work acceptably. But all this took time, the very item the production engineers had hoped to save by the use of dies.

Still more troublesome was the matter of spring-back. The Ford engineers discovered that aluminium would not retain the exact shape given it by a forming die. To correct this they had to design a sequence of two or more dies to perform deep draws in successive steps where a single pass would have sufficed with steel. Ford ended up by making 29,000 dies, of which only about one-half were actually used. Moreover, about 2,400 of these had to be reworked, some of them repeatedly, before they were satisfactory.

The use of dies proved disappointing in other respects. Where the airframe builders laboriously drilled holes one at a time, or sometimes faster with gang drills, the automotive men planned to punch out the holes for an entire skin section with one pass of a press. The idea was alluring but it did not always work. Aluminium skin sheets showed a distressing propensity for stretching irregularly and thereby failing to mate properly when they reached the point of assembly.

Finally, the inevitable problem of design change militated against the use of dies. One of the major economies anticipated was the long production run with semi-skilled labour. A single tool setting, the engineers hoped, would suffice for the full number of items on any given contract. The high frequency of design change destroyed this concept entirely. The planners found it was unwise to stamp out more than a sixty-day supply of any part since beyond that point a design change could make scrap losses extremely serious. Furthermore, it often turned out that shortages of critical materials made even a limited sixty-day run impossible.

Although the Ford engineers made extensive use of tooling to speed up the fabrication of individual parts, it was in jigging up for final assembly that they carried their ideas to the ultimate. They built fixtures for every single assembly operation.

One of the most impressive was the massive device used in assembling the B-24 wing centre-section. A pre-drilled and pre-cut aluminium sheet to make the upper wing surface was placed in Fixture No. 1, where rivets were inserted in the holes. Fixture No. 2 then closed down on this skin to hold it in place. Fixture No. 3 held splicer bars in position while they were riveted to the skin. The next two fixtures passed up stringers and locked them into place while they too were riveted to the skin, and a sixth fixture held both stringers and splicers while they were riveted to each other. Finally, an overhead conveyor moved in, lifted out the finished component, and carried it off to the assembly line. Eventually Willow Run had seven banks of these wing assembly fixtures, each bank holding five separate wings. Thus thirty-five individual wing centre-sections were under construction simultaneously. Each fixture was 60 ft long and 15 ft high; a mass of cast iron and steel. They were costly and required months to build, but once in operation they needed only one-sixth the labour required by conventional methods of aircraft construction.

If the centre-section jig was impressive, even more so was the huge milling apparatus installed to machine the finished subassembly. Using this special machine tool, a seven-man crew did some forty-two machining operations in 3½ man-hours. With conventional tools the same job would have required 500 man-hours. But an even greater advantage of this tool was the gain in accuracy that it permitted. With the device, all four engine mounts were milled and drilled simultaneously in perfect alignment. So too were the main landing gear bearing holes. At one stroke, Ford had mastered the immensely taxing problem of alignment.

The whole Ford assembly plan was bold in conception; nevertheless, it too suffered from the fundamental error of premise that underlay so much of the Ford experiment, i.e.—that the B-24 design would be reasonably stable. And since frequent changes in design characterised the whole production life of the B-24, revisions in the elaborate tooling had to be made just as frequently. Altogether, the Ford production men built 21,000 jigs and fixtures, but only about 11,000 were finally put to use.

Nevertheless, the realisation that the B-24 design could never be frozen came hard to the Ford engineers. Only gradually did they begin to understand that design change was a perfectly normal attribute of military aircraft. This fact resulted in the compromise with the Air Force which allowed Ford to produce a 'semi-frozen' design—the B-24E—long beyond the time when this variant was a viable combat machine. (As noted previously, the entire production run of B-24E's was placed in an obsolescent category the day after the last one was produced.) This compromise, in effect, allowed Ford to reorient its production thinking while producing an abundant supply of B-24 trainers for the Air Force.

There is an ironic postscript to the Ford story. When North American joined the Liberator pool, their engineers made almost the same criticisms of Ford that Ford was making of Consolidated: production in Dallas was hindered by faulty liaison. The drawings provided by Ford arrived in unsatisfactory condition. As experienced airframe builders, the North American engineers fully expected changes in design, but they were disturbed by the laggard pace at which the 180,000-odd change notifications had cleared through Ford during the first few months of production. And just as the Ford staff before them had redone all the drawings received from Consolidated, the North American engineers finally decided to redraw all the prints sent to them from Willow Run.

The View from the Ground

Most World War II aircraft spent most of their lives not flying, and for every crew that took one aloft there were many times that number of men concerned with making the trip possible. How did these men—the mechanics, the crew chiefs and the line chiefs—feel about the Liberator? Here is a composite commentary, distilled from offerings of a representative sample of those truly unsung heroes.

'I worked on both the B-17 and B-24 and I liked the B-24 better from the standpoint of maintenance. It was not as simple, but it was cleaner and it had more space to move around and work in. It was more modern in concept and the various systems were of more modern design. The engines were accessible; you could butterfly most of the cowlings without too much trouble and you could easily get to most of the sparkplugs and that sort of thing. The back end of the nacelle was pretty accessible too, because it had large panels that were easy and quick to remove. The Pratt & Whitney R-1830 was a much cleaner running engine than the Wright R-1820 that the B-17 had. The Wright engines were well known as oil slingers and made a mess of the cell and the areas around them. As far as the airframes themselves, the B-17 was an old design coming off the drawing board in the thirties, I believe, and as a result it was not a flush riveted, smooth skinned affair. It had a lot of beef to it, though. The Liberator, of course, had the new Davis wing which was a high lift aerofoil and the ship was considerably faster. This was one of the features I think a lot of crews liked; the fact that you could pour the coal to it and it would move out.'

Top: Willow Run wing section jig. /Ford

Above: This 15th AF B-24J-CF received a direct flak burst in the waist section. The shell did not explode until it hit the roof inside the bomber, wiping out the two waist guns and severing the elevator and rudder cables. The pilot brought the plane back on engines only. /USAF

'The Fortress was a tight aeroplane compared to the Liberator inasmuch as the nose gunner and tail gunner were in very confined areas and the bomb bay was smaller—it was not as long—although the loads carried usually were about equal. As an older, tail wheel type aeroplane it was gawky looking on the ground, with legs sort of spread out and all that. It gave most of the fellows the impression that it was not the latest thing, even though it looked somewhat nicer in the air than the Liberator did.'

'As I said before, the B-17 was a rugged aeroplane, structure-wise. It was beefy, being of an older school of engineering design. But don't let anybody tell you that the B-24 couldn't take a hit and come back, too. You wouldn't believe some of the wrecks that those guys flew home.'

'Running-up a B-24 was a one-man operation inasmuch as it had electric boost pumps in the fuel system and everything was grouped in an area on the co-pilot's side of the cockpit, where it only required one man to start the thing, check it out and do a complete run-up. The B-17 required two people to start the engines, since it had an older, hand primer pump with a selector and you had to either be 8 ft tall with a 10 ft reach or have two people to start it. Even with two it was a pain in the neck. The whole arrangement up on the flight deck of the B-24 was much better. It was quite roomy, and of course it sat on a horizontal plane whereas the B-17 was a tail-wheeling deal. So the Liberator was a lot more comfortable up front, and a run-up was usually accomplished without too much trouble.

'On the subject of check-outs, the B-24 had another very good feature. It had an auxiliary power unit, or APU, mounted in the aeroplane which was used for starting and pre-flight checking for the crews, running the turrets, and checking the radio gear. It was a nice unit, it was handy, and it started off the ship's batteries. It took a mixture of gas and oil similar to today's lawn mowers. I recall we had to change one every once in a while but it was easily removed and replaced. The B-17 didn't have a built-in APU like the Liberator did. I recall we used the same identical auxiliary power units but they were mounted on small carts and had a long extension cord. We used to have to hand crank them to get them running and then plug the cord into the ship, which was quite a pain because the carts were always kept outdoors and even though we covered them with canvas it was hard to keep them in the best of condition.'

'The B-17 was a dirty aeroplane compared to the Liberator—those engines sure threw a lot of oil. Because it was an earlier design, some of the systems were not as modern; the supercharger system, as I recall, was oil-pressure driven and it was a real filthy job changing one, whereas the Liberator had an electrically operated and controlled supercharger. The rest of the B-17, of course, was all electric—landing gear, flaps, and so forth. We always used to say, and it still is said in industry today, that if it's Boeing it's all electric and if it's Douglas, it's all hydraulic. In those days Consolidated also meant all hydraulic. It was either a fault or an advantage, I don't really know. We had trouble with the B-17 landing gear motors jamming. As a result we had gear trouble on the B-17's which we never had with the Liberator. It was just one of those things you like or dislike. Both systems had their faults but I liked the hydraulic better. It was sure less troublesome.'

'The B-24 engine was easier to replace. Our maintenance squadrons were set up for a four-point change, which meant the engine nacelle pack would be replaced as a unit and the four points referred to the fact that there were four bolts attaching it to the airframe. Sometimes we had to make engine changes on an eight-point basis, though, which meant removing the engine right from its mount. Still, that wasn't so bad. It was attached at eight points with rubber shock mounts. We removed the engine from the mount, after removing the carburettor and associated lines and equipment, and took the engine off of the pack and got a new engine out of a box.'

Top: More work for the weary—a 30th Bomb Group B-24M after an aborted take off from Wheeler Field, Oahu.—US Army

Below. Ground work was hard all over. / Via J. Scutts

'Most of the crews I worked with had flown both B-17's and Liberators and they seem to have preferred the Liberator. They liked the fact that it was a faster aeroplane—it would get out and over the target and be gone if it had to. Of course the Liberator had some bad features, too. One of the worst—not as far as maintenance was concerned, but as far as the aircraft in general was concerned—was the auxiliary hydraulic pump mounted in the bomb bay on the right side. This pump was actuated by a pressure switch, and if you had the darn thing actuated and you were taxiing around prior to take-off, it was forever cycling on and off. It was electric-motor driven and the brushes were exposed to the atmosphere and of course when there is an electric motor running with brushes there are sparks. This would have been all right if it weren't for the fact that, with all those fuel cells and interconnections in the wing, the Liberator was usually leaking a small amount of fuel which, because of the dihedral angle of the wing, would seep towards the centre and run into the bomb bay. At that point, with the hydraulic pump cycling on and off without adequate ventilation, the ship was a flying bomb. Whenever I went on cross-country hops, I always shut the damn thing off or insisted that the crew shut it off. It was not a good feature, and I cannot see to this day how they could design something like that, out in the open without any proper guard over it and no outside venting.

'Another feature about the aeroplane that you could say was a fault was that it was not balanced too well on the ground. Without a load it was kind of tail heavy and we always used to have an ammunition box positioned under the skid in the rear so the darn thing wouldn't touch the ground when we hopped on board through the rear door. Also, because there wasn't too much weight on the nose wheel it was hard to taxi using only minimum power because you couldn't get the nose wheel down firm on the ground. You had to draw quite a bit of power from the engines so that you could get the nose down and then keep it there with the engines and the brakes. Usually we had all the flak suits and all the gear the crews would leave on board up front to help balance it out.'

'A bad point about the B-24 was the fact that the system of priming the engines was highly susceptible to leaks, and as a result we had quite a few fires on starting engines in the morning. The primer system incorporated very small pieces of tubing, about 1/8 in. outside diameter. The lines which primed the top five cylinders all came together around the back end of the number one cyclinder. It looked sort of like a spider, and that's what we called it. Now, if one of those lines got broken by vibration or something, when you pushed the primer switch you started spraying fuel directly on the outside of the engine and around the cyclinder and, if conditions were right, you got a fire. The fire wasn't too serious in itself, because you could stop it by cutting off the flow of gas, but you had to do it quickly because all of the lines in the back end of the accessory section were made up from plain low-pressure rubber tubing and once the fire broke in there it would destroy them quickly. Of course, you can't blame these incidents on the aeroplane itself—it was a bad design feature of the engine.'

'There was a lot of sentiment about the aeroplanes among the men who worked on them. They felt that they were their own—sort of the way you treasure a new car today. They kept them as clean as possible and took a lot of pride in their appearance and performance. It sure was a sad day when one didn't come back.'

NOTES

1 The Model 31, never destined for volume production, later spawned an interesting offspring: the Model 38. Marrying the wing R-3350 engine package and tail surfaces from the Model 31 to a standard B-24D fuselage with modified armament, CAC offered the result to the US Navy as a land-based search attack

aircraft. Although it did not progress beyond the study stage, the Model 38 was, in effect, a twin-engined B-24.

2 The original French contract was not signed until 4 June 1940, only two weeks before France capitulated.

3 For example, it is relatively certain that the XB-24 evolved from design proposal LB-16. The XB-32 probably stemmed from design proposal LB-25. The twin-engined Liberator for the USN, previously mentioned, was design proposal LB-29.

4 Throughout 1941, however, constant power plant problems were encountered with the -41 engines and on 26 January 1942 the Material Division at Wright Field approved the replacement of these engines in the XB-24B with R-1830-43's.

5 When His Majesty's Government took over the French contract the number of aircraft on this order was reduced from 175 to 159. These 159 aircraft, plus the six already on order under Contract A-5068, were allocated RAF serials AL503 through AL667. The last twenty-six numbers of this allocation were cancelled and new serials allocated when the two different (and earlier) Liberator models were later substituted under the previously noted US-UK contract exchange. This explains why the LB-30A's and LIB I's had higher serials than the LIB II's. For additional contract details, see Appendix A.

6 Because of the possibility of sabotage, this accident was thoroughly investigated and the results exhaustively documented. Photographs show that the jammed bolt was, without question, the cause of the accident.

7 Contract ac 12464 originally called for seven YB-24's to be given serials 39-618 through 39-687. Since the exchange delayed the actual purchase of these aircraft until Fiscal Year 1940, the earlier serials were changed to 40-696 through 40-702.

8 This aircraft was the flagship of Force Aquila, the forerunner of the Tenth Air Force in India. Flown by Colonel C. V. Hanes, 40-698 reached India on 7 April 1942. As far as is known this was the first USAAF B-24D to arrive in an operational theatre.

9 As an interesting sidelight, CAC applied for and was granted a Registered Trade Mark (No. 401,692) for a circular insignia which contained a drawing of a B-24 in flight and the words CONSOLIDATED LIBERATOR.

10 The USAAF assigned two letter designations to indicate the name and plant of the manufacturer of an aircraft. These were included as a part of the stencilled data on the left side of the fuselage nose. These designators will be used throughout the book as a convenient form of shorthand and are defined as follows:

CO: Consolidated, San Diego
CF: Consolidated, Fort Worth
FO: Ford, Willow Run
DT: Douglas, Tulsa
NT: North American, Dallas

Consolidated did not use a true block system until the advent of the B-24J-1-CO. Prior to B-24D 41-23640 no 'batch' system at all was employed. During the remainder of D production, aircraft were produced in what were called 'series'. These series numbers resembled block numbers and later came to be treated as such. This practice has been followed in this book. However, it will be noted that many major changes to the B-24D took place in the middle of a series, which would not have been the case if a true block system had been in use.

11 More accurately, the centre fuselage was built around the wing. This saved juncture weight through the elimination of heavy attachment fittings and fuselage torsional stiffness was increased.

12 On Liberators prior to serial 41-23640 this figure was 2·40 sq ft.

13 This incident is probably traceable to the interaction of the airflow on the vertical tail surfaces if the cowl flaps on the No. 2 and No. 3 engines were opened under certain flight conditions. This phenomena was not generally known at the time.

14 The fact that B-24E's were used in combat was, to the author, one of the most surprising facts uncovered during a dozen years of research on the B24.

15 The idea of a B-24 being shipped over 1,000 miles via 'covered wagon' a few weeks before its appearance over Berlin invokes an undefinable but strangely anachronistic feeling—the significance of which, if any, has thus far eluded the author!

16 Joe Famme, who succeeded Frank Fink as B-24 Project Engineer, has written that the 'need for nose turrets was stressed in the early days of operations of both the B-17 and B-24 airplanes by the Eighth Air Force. The Bendix chin turret, as originally designed for the XB-41 airplane and later released for B24 production, was cancelled by the Air Technical Service Command and the production and design of the Bendix turret was turned over to Boeing for use on the B-17. The only turrets remaining for use on the B-24 were the MPC-5 . . . and the Emerson Model 127.'

17 Vultee Aircraft Corporation had acquired working control of Consolidated in late 1941, and on 17 March 1943 the two companies merged to become Consolidated Vultee Aircraft Corporation (CVAC), later contracted to Convair.

18 This was later revised, of course.

19 MCR was, in a way, a double acronym. To Wright Field it meant Master Change Record, a procedure established to identify and code in a uniform manner the major engineering changes in each aircraft model. To the builders it came to stand for Manufacturer's Change Request, the detailed description of a particular modification plus the schedule for introducing the change on the production line.

20 There is some irony here. CAC's original motivation for looking into the single tail was to obtain these increases in field of fire for the gunners. The expected improvements in stability and control were considered to be bonus benefits.

21 The XPB4Y-2's, which were initially fitted with a smaller vertical fin and rudder, suffered instability problems. To remedy the situation in the shortest possible time Laddon proposed outboard stabiliser fins such as had been applied to the XPB2Y-1 under similar circumstances. Although both inboard and endplate ('Zulu Warshield') designs were examined, both were rejected by the Navy, which was willing to accept delayed deliveries in order to get the taller tail.

22 Canadian 'Lend Lease' aircraft were actually paid for in cash throughout the war.

23 Costs came down also as production mounted. Each B-24A cost $341,960; a typical B-24J cost $210,943. These totals can be broken down as follows:

	B-24A	*B-24J*
Airframe	$225,125	$115,188
Engines	53,840	30,600
Gov't Furnished Equipment (GFE)		
Propellers	11,460	4,220
Radio gear	9,520	7,092
Ordnance	3,407	3,152
GFE for airframe	36,547	48,235
GFE for engines	2,061	2,456
	$341,960	$210,943

24 The balance of this section is largely based on *Buying Aircraft: Materiel Procurement for the Army Air Forces* by I. B. Holley, Jr., published by the Department of the Army in 1964. Much of the material previously appeared in the 1945 Air Technical Service Command study *B-24 Construction and Production Analysis—Ford, Willow Run.*

10 February 1945

'The manpower problem is extremely acute. Daily we are compelled to resort to every conceivable device to find enough manpower to keep the air effort rolling here. It is understandable, therefore, that General Arnold was gravely concerned when, during a recent inspection, he found that our Mod Centers were performing 500-odd modifications on the B-24 alone. He at once directed that the number of mods performed on the B-24 would be held to a minimum, and that B-24's for all theatres would be standardized to the greatest extent possible. . .

'This decision must stick, for we must standardize and it is of the utmost importance that we do not change our directive to the manufacturers as frequently as we have done in the past.'

BARNEY M. GILES
Lieutenant General, USA
Chief of Air Staff, Army Air Forces

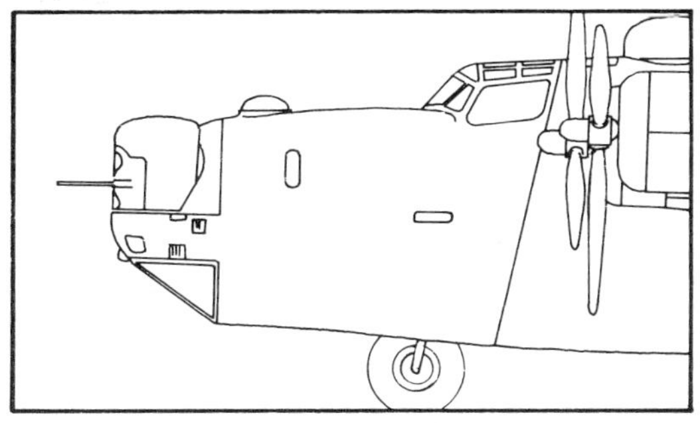

II Modifications and Conversions

General

Because of the versatility of the B-24, modifications to the aircraft were extremely varied and often complicated. Most military aircraft were procured on the basis of their adaptability for a particular mission or type of warfare. The B-24 was not a one-theatre or single-mission aircraft, but a USAAF work-horse, fighting the war on all fronts. It was this wide use of the Liberator that led to most of the modification difficulties which came to be associated with the production programme during 1944, when over fifty separate and distinct versions of the Liberator were being produced simultaneously by the modification centres.

The military requirements for effective combat use of the Liberator differed widely. The USAAF units operating in Europe worked closely with the British, so that radio, radar and other equipment had to be different from aircraft going to Pacific theatres where missions were conducted over large water areas working in close connection with the Navy and widely separated shore stations. Radar equipment for European and Mediterranean operations was needed to permit the bombing of land targets from a considerable altitude, while Pacific radar bombing was usually accomplished from a lower altitude using the low-altitude bombing attachment against shipping. Cold climates required different equipment than hot ones.

Considering such differences in theatre requirements, it can be seen why it was impossible for the factory production lines to turn out a finished product, equipped with all the necessary gadgetry, special radar and other devices required by theatre commanders to perform desired combat missions. It was common practice for Washington to grant most requests of fighting commanders, feeling that they were the people that were actually fighting the war and knew what should be done to the aircraft. This policy resulted in such a large number of changes that contractors could not possibly incorporate them all and still maintain production schedules. Regardless of how minor a change might appear, many long processes were involved to design, develop, test, standardise, procure materials, produce tooling and fabricate each new part. A change also meant the changing of affected production design and shop drawings, re-instructing installation personnel and changing all affected records including those for re-scheduling spares and service installation kits. Normally fifteen to thirty changes were in work at one time with a backlog of seventy to 150 changes pending. This condition was aggravated by the lack of positive and effective USAAF control of priorities for changes, since each contractor had a tendency to apply his own priority to suit the convenience of his facility with only secondary regard for the urgency on the change. This situation tended to aggravate the already serious problem of non-standardisation between the B-24's produced at the various manufacturing sites.

Major B-24 modifications were accomplished by seven primary modification centres, assisted when possible by the prime assembly plants (see Appendix N). Arrangements were made to have such items deleted in production which might save the modification centre time and work; for example, H2X aircraft required the installation of the AN/APS-15 radar equipment in the position normally occupied by the lower ball turret. After the H2X installation had become standardised, Ford deleted the lower ball on sufficient aircraft which were then modified by the St Paul modification facility to fill H2X requirements.

Modifications accomplished on a production line basis required a minimum of 500 up to a maximum of 5,000 or 6,000 man-hours for some of the more complicated radar and photographic installations. The first installation, or mock-up, often ran to a much higher figure.

During July 1944, B-24's were being produced in such large quantities that modification centres were unable to accommodate all of them. (Another factor was that a considerable amount of modification space previously devoted exclusively to the B-24 was taken over by the higher-priority B-29.) In order for production to continue, it was necessary to set up storage pools at different points in the United States where the aircraft could be retained until the mod centres could again accommodate production. This pool of aircraft continued to grow until over *900* new B-24's were in storage on 1 January 1945. By VJ-Day this backlog had been reduced somewhat, but even so over 400 unused B-24's still remained at storage pool locations when the war ended.

One of the first USAAF wartime Liberator modifications (the Tech Order was dated 30 December 1941) was to install a 'top gun assembly' on its single B-24 (nee YB-24 40-702) and nine B-24A's. In essence, this involved placement of a ring assembly above the top hatch to accept a single flexible machine gun which could be manned by a gunner standing below.

The last three B-24C's, 40-2384/2386, were used by Wright Field to test a variety of proposed Liberator modifications including, among others, a slightly larger (46 in. *v.* 44 in.) retractable ball turret and various other armament changes. Number 40-2386, in fact, received a full-house gun treatment by General Electric consisting of an upper, tail and non-retracting lower turret, all of which were electrically operated and sighted remotely. This was the fire control system GE was proposing for the B-29 and, after additional testing, some 160 B-24E's are reported to have been similarly modified during January 1944 at Mobile, Alabama, and Warner-Robbins, Georgia, to serve as trainers for B-29 gunners. Some B-24L's were later added to this inventory.

While on the subject of armament modifications, the XB-41 must at least be mentioned. Converted from B-24D-CO 41-11822 to evaluate possible use of the design as a bomber escort, the ship, which had been assigned to San Diego for test purposes and never officially delivered to the USAAF, was transformed to the XB-41 at Fort Worth. It was given an extra Martin turret topside (like the US LB-30's), a Bendix chin turret and an extra gun at each waist hatch for a total of fifteen ·50 calibre machine guns if we count the good old tunnel gun. The total ammunition carried aloft for this battery was 12,420 rounds. The regular (forward) Martin turret installation was modified so that it could be raised during use, thus increasing its field of fire, and lowered, when not needed, to gain an extra 3 mph. The twin waist guns were power-assisted and, as a comparison test feature, the left waist was fitted with a Plexiglas blister while the right waist guns were not enclosed. The single XB-41 was then delivered (on 29 January 1943) and was tested by the Air Force Proving Ground Command at Eglin Field. After a test programme which lasted most of the spring and summer, Eglin first approved the conversion (16 March) and then rejected it (21 October) as operationally unsuitable. It was subsequently recommended that the XB-41 be used as a test bed for the development of a four-gun nose turret, but no record of this project has been located. Later the XB-41 was redesignated a TB-24D and was relegated to a Class 26 (Instructional Airframe) on 3 September 1944. The B-17 counterpart of the XB-41, the YB-40, was similarly judged operationally unsuitable during a brief combat

Top: Often at war with the weather, the B-24D served in every theatre of operations. Photo shows radar equipped Liberators from the 28th Composite Group in an Aleutian snow storm. /USAF

Above: The B-24L modified for B-29 gunnery training. /W. T. Larkins

trial of a number of these aircraft in the European theatre.

An early modification plan aimed at helping solve the logistics of world-wide war was to strip the interior of a fleet of B-24D's so that each could carry two knocked-down P-39 Airacobras to the South Pacific. Although it appeared feasible, this experiment did not progress beyond a single test aeroplane.

A representative sample of the many other B-24D modifications might include 41-11707, which was used at Wright Field to test the conversion to wide-blade or 'paddle' props; 41-11804, which was altered to assess the value of certain production changes requested by the Seventh Air Force; 41-24185, which was used to test an early nose and ball turret configuration; 41-11893, which was fitted at Fort Worth with a trial 37 mm cannon in the nose; 42-40200, which served as the test aircraft for Project Yehudi (camouflage by aircraft self-illumination); 42-40400, which was modified to a training ship with special electronic equipment; and 42-41003, which was used at San Diego for tests with thermal anti-icing features.

A similar representation from UK Liberators might include AL505 and AL546, which were used for turret, bombing system and flame damping trials. One Liberator II was also fitted for test with a 400 lb smoke screen dispenser. BZ801 was used for handling trials with retractable RP equipment fitted, while BZ967/G was modified to test the feasibility of carrying and dropping torpedoes. The RAF was also concerned with the poor ditching characteristics of the Liberator, and some testing was carried out to find appropriate solutions to the problem. One of these was particularly ingenious. A small retractable hydro-ski was positioned under the nose of the aircraft. In the ditching configuration, wheels up and hydro-ski down, the latter was expected to provide enough 'lift' to keep the nose and vulnerable bomb bay structure off the water until speed had been reduced to the point where it was safe for the fuselage to settle in. While towing tank model tests were successful, the doubtful utility of the system in rough water and the pressing business of war took priority over modifying the comparatively few Liberators then in British service.

Later, in 1943, USAAF ditching experience began to cause no little concern over the exceedingly small percentage of crews rescued and the problem was given considerably more study. Two approaches were taken. The first was to cut additional escape hatches in the fuselage and, by interior rearrangement, position as many of the crew as possible on the deck over the bomb bay during impact. Two canvas 'ditching belts', or slings, were added to protect a total of six men at this station while the remainder of the crew was instructed to stay in the engineer-radio compartment or on the flight deck. Trial modifications of this type were carried out by Lockheed Overseas Corp. and Scottish Aviation Ltd. The second modification was to reinforce the relatively light construction of the bomb bay doors to prevent their collapse upon impact with the water. This feature eventually was introduced on production Liberators in the form of four pairs of detachable 'bomb bay door stiffeners' or 'ditching ribs', which were carried on board the aircraft and slipped into place by the crew prior to ditching. Even with improvements, the Liberator remained a poor ditching risk. Eighth Air Force statistics for 1943-5, for example, show that a B-17 crewman's chances of surviving a ditching were 37·9% while the equivalent B-24 figure was 26·5%.

Because of its early use in antisubmarine work, the Liberator was associated with airborne radar almost from the beginning. The British installed their long-wave Air-to-Surface-Vessel (ASV) radar on many of the Liberator I's they received during the first half of 1941. Samples of the ASV were taken to the US during the same year, flight tested in a B-17D in July, and accepted for US use as the SCR-521 in August. Contracts were let in October and by July 1942 there were over fifty of these sets in use. However, certain technical shortcomings of the ASV, plus its bulky, drag-producing antennae, had led to the development of the ASV-10 (SCR-517), a microwave set using a much cleaner radome antenna configuration. After Wright Field mock-ups had been prepared for installations on several

Top: The XB-41, shown here on 2 May 1943 while at Eglin Field for operational suitability testing. Five twin-fifty power turrets plus hand-held twin fifties in each waist hatch.

Above: The XB-41 left waist gun installation as delivered. The curved Plexiglas bubble was found to cause severe distortion and was removed early in the test programme. / USAAF via Tom Davidson

Left: Detailed view of flame dampers installed on AL505 for test, 1942.

Below: Liberator II AL504, Prime Minister Churchill's *Commando*, was returned to the Consolidated Modificaiton Center in Tucson in 1944 where it was extensively modified, including a fuselage stretch and single tail. Since Churchill's last flight in *Commando* was on 7 February 1943, he never occupied the aircraft while it was in this configuration.

Right: A Liberator I, AM910, fitted with early ASV radar and cannon. Ground test firing of the latter caved in the bomb bay doors but this problem was not encountered in the air. Photo taken in July 1941. /USAF

Below right: Later in the war the 868th Squadron experimented with drop tanks to increase the range of its Liberators, which by that time were of the natural metal finish variety. Photo shows tank shortly after release.

AM910

450395

types of aircraft, DUMBO I, Liberator II AL507, was flown to the UK in March 1942 with the first actual installation. (DUMBO II was AL593. The two DUMBO's were the last of the repossessed Liberator II's to be returned to UK control; AL507 returned to the USA in June 1942 and did not reach the UK for keeps until 15 October, while AL593 was taken on charge at Montreal on 12 January 1943. The full story on repossessions is described later under Operations.) Tests carried out from Northern Ireland by 120 Squadron proved highly successful and additional SCR-517 sets were rushed to completion, installed in Liberators and delivered between August and December 1942. These aircraft, LV336/346, included US serials 41-1087, 093, 096, 097, 107, 108, 111, 114, 122, 124 and 127.

On the other side of the world a surprise was brewing for the Japanese also. This was the low-level, blind bombing SB-24 or 'Snooper'. Equipped with a special radar scope for the bombardier and a computer that translated and transmitted the information on the scope to the bomb release controls, plus other special equipment, the black-painted SB-24's were capable of entirely blind bombing of moving targets from low altitudes under conditions of total darkness or complete fog. SB-24 engines were fitted with flame dampers which turned out to be so successful that, at night, these aircraft were invisible and practically noiseless beyond half a mile.

A special detachment of ten SB-24D's left San Francisco on 10 August 1943 and arrived on Guadalcanal on 22 August. Initially attached to the 5th Bomb Group, the 'Snoopers' began strike operations five days after arrival. In ten weeks 111 individual missions were flown, averaging over eleven hours each. Results were, to say the very least, beyond expectations. The ten Liberators made ninety-four night runs on ship targets, at an average altitude of 1,500 ft, with a *direct hit* average of 24%. Two SB-24's were lost, only one of which was probably the result of enemy action.

In January 1944 the Snoopers won their independence and became the 868th Separate Squadron, reporting directly to Thirteenth Air Force. Their continued successes (and the course of the war) soon made low-level night shipping targets few and far between, and the 868th moved on to other special missions. However, it should be noted here that well into 1945 the first of the SB-24's, numbered 42-40639 and originally nicknamed *The Lady Margaret,* was still carried as 'ready' on the status board of the 868th.

Because of the hurry to get the ten original SB-24's into action, they by-passed a major modification programme that had been laid on at the Hawaiian Air Depot early in 1943 to modify B-24's assigned to Pacific theatres. Some twenty-five major changes were involved, including installation of nose turrets, retractable ball turrets, and replacement of tail turrets with hand-held twin ·50's. In addition, nose turret conversion kits were supplied for Australian-based Liberators which were installed on these aircraft at Brisbane. The depot also handled nose structural modifications on Navy PB 4Y-1's so that they could accept the Emerson 128 ball turret then being installed as nose armament by that service.

Another nose turret modification programme was active for a limited time at Oklahoma City during the early summer of 1943. Although B-24D's were the normal starting point, there is a persistent rumour that a small number of B-24E's were included in this programme. While this story remains unconfirmed at this writing, it would go a long way towards explaining not only why a few B-24E's made it overseas with operational units, but also why they were considered fit to be flown in combat as late as mid-1944.

During many of these early Liberator modifications it became apparent that the relocation of weight associated therewith, particularly in the case of tail turret removal and/or the installation of special equipment, shifted the centre of gravity to the extent that at the end of a flight the c.g. could fall in front of the safe forward limit for landing. In these aircraft all personnel except the pilots and radio operator were directed to move to a position just forward of the waist windows during the landing operation.

Late in the war the 'ground launched

An Oklahoma City nose mod in service with the 479th Anti-Submarine Group at St. Eval, England, in 1943. /USAF

Above: A reworked B-24D-1-CF 'mother ship' with two Culver PQ-14 radio controlled target planes. Note that all armament, including the turrets themselves, have been removed from the Liberator and the exterior stripped to bare metal. Aircraft are from the 17th Tow Target Squadron based at Wheeler Field, Hawaii. Photograph was taken in April 1945. / USAF

Below: The XB-24F. Shown here at the Ames Aeronautical Laboratory of the NACA (Moffett Field, California) where conversion to hot-air de-icing system was performed in March, 1943. This B-24D-CO, 41-11678, was delivered 9 April 1942 and loaned to Moffett for test purposes. Actual change in designation to XB-24F was made while the ship was at San Diego factory the following month. Aircraft was later loaned to Northwest Airlines Mod Center at St. Paul and, still later, to AMC at Tinker AAB. It was scrapped there on 17 April 1947. / NACA

Top right: The NACA Langley test aircraft goes in . . .

Centre right: . . . comes to rest with a typical ditching fracture at Station 4.0. . .

Right: . . . and is recovered. / General Dynamics

controllable bomb' idea was resurrected for use during the final assault on Japan's home islands. An undetermined number of B-24 war-wearies were thus converted to BQ-8's under this programme but none is known to have been employed operationally.

As it did with other warplanes, the US National Advisory Committee for Aeronautics (NACA) ran extensive tests on the Liberator. It was the NACA Ames facility at Moffett Field, California, that fathered the sole XB-24F, a standard B-24D which was modified for hot air de-icing. The NACA Lewis installation near Cleveland, Ohio, used B-24D 42-40223 as their experimental Liberator for two years while running various tests including the use of fuel additives to increase performance. Number 44-41986, a B-24M, was also a long-time Lewis resident. From November 1945 to July 1949 it served as a test bed for icing research on jet engines, antennae, radomes and windshields. Perhaps the most dramatic NACA testing, however, was done by the Langley, Virgina, facility. In August of 1944 it acquired B-24D's 41-23679, 41-23878 and 42-40198 to run full-scale ditching tests. Although only one was actually carried out, it was well-advertised and was without doubt the most well-watched Liberator water landing in history.

Throughout its life the XB-24B continued to serve as a test bed for Liberator modifications. One of the most interesting of these was the installation of full main wheel fairings to investigate the effect such streamlining would have on the aircraft's speed and rate of climb. During March 1942 a dummy modification was made, in which the gear was left in the down position and a non-operating fairing of the proposed configuration was applied over the wheel and strut openings in the wing. When flight tests of this configuration indicated an increase in speed of 9·5 mph with 100% power at 25,000 ft, it was decided to proceed with the installation of actual working fairings. These consisted of three parts: the fixed after fairing, a fairing piece attached to the landing gear, and a hydraulically operated door hinged on the wing at the outboard side of the wheel well.

As a result of other commitments the working fairings were not flight tested until the period 16 April—21 May 1943. Because of the difference in pressure sealing between the dummy fairings and the actual working fairings, the speed increase was reduced to 7·0 mph under similar flight conditions. Even so, the XB-24B turned in a respectable 319 mph in level flight with its neatly enclosed gear.

Somewhat later, *Old Gran'pappy,* as the XB-24B was affectionately known around the San Diego plant, was outfitted with full-span flaps (being considered for the B-36) and at one time also appeared with a gloved wing, presumably in an attempt to decrease what had become one of the Liberator's main faults—high wing loading. It was in association with the gloved wing experiment that the XB-24B was subjected to a series of dive tests to measure pressure profiles behind the wing. During one of these dives a *true airspeed of 465 mph* was recorded.[25]

During 1944 the USAAF contracted with Convair to convert the XB-24B into an executive transport, and by the end of the year this conversion had been completed. Sporting R-1830-65 engines, large picture windows in the fuselage and a plush interior, *Gran'pappy* was delivered to the 6th Ferry Group on 25 April 1945. Operating out of Washington National Airport, the ship transported VIP's (many on overseas flights) until March 1946. On 30 April the XB-24B was reported at Brookley AAB in Alabama and on 12 May 1946 it was dropped from USAAF inventory and reduced to salvage.

Consolidated's long association with Pratt & Whitney was not without its periodic squabbles. On one occasion in 1943 Laddon, who could be stubborn at times, vowed he would re-engine the B-24 with the Wright R-1820 if Pratt & Whitney didn't give in on the particular point (now long forgotten) which was at issue at the time. To back up his threat he had a single R-1820 installed in the No. 2 position (port inboard) of a B-24 which retained its regular P&W R-1830's at the other positions. According to CAC test pilot Phil Prophet, to whom befell the dubious honour of flight testing the result, 'With that left

Top: This is the XB-24B flight deck layout as it looked on 19 April 1945. / General Dynamics

Above: Until late in the war the original Liberator served with the Flight Test Division of Consolidated as an experimental aircraft testing modifications proposed for production Libs and later for the B-32 programme. Note the late engines, extended nose, new cockpit enclosure, etc. This aircraft, the granddaddy of them all, was finally 'reclaimed' (ie, broken up for scrap) in 1946 with 3000 hours on the airframe. / General Dynamics

View of C-87 from below. / Consolidated

inboard turning 2,550 rpm on take-off and all the others putting out 2,700, you just wouldn't believe the racket that beast made.' In the end, of course, tempers cooled, logic prevailed and the hybrid test aircraft reverted to form. For a time, however, it was—at least as far as Laddon was concerned—the only B-24 flying on three Wrongs and a Wright.

On another occasion the Consolidated Flight Test Department was given a B-24 with the bomb bays completely sealed and taped shut in order to determine just how much drag was present in the normal configuration. Murphy's Law was in full effect that day and a transfer valve, normally used to transfer fuel from the auxiliary bomb bay tanks, was inadvertently left open. Since there *were* no auxiliary tanks installed in the aircraft being tested, gasoline from the wing cells literally poured down into the bomb bays—where it couldn't get out because of their sealed condition. After an abbreviated flight the Liberator touched down at Lindbergh Field with about 2 ft of raw 100 octane sloshing around in the bays and the flight crew muttering about the strange way that some people build bathtubs and then forget to put in the drain plugs.

In the operational theatres, the search for perfection was never-ending. As late as January 1945 the following modifications were being performed on all B-24's arriving in England for the Eighth Air Force:

1. A complete nose turret seal installed to reduce the draught which entered the aircraft around the turret. This seal was activated by wind pressure.
2. Armoured glass installed on the forward windshields of the pilot's compartment. The free air temperature guage was relocated to fit through the window frame at a point close to the original installation on the Plexiglas window.
3. Vertical attaching brace added for the flak curtain; pilot's arm rests and attaching lugs removed to allow easier egress from the pilot's station.
4. Bomb bay lock release mechanism replaced by a system which would allow the doors to creep the maximum safe amount, 6 in., before locking the bomb release either mechanically or electrically.
5. Life raft release system changed to provide greater leverage at the internal point of release and an external release added just outside the rear escape hatch.
6. Top turret armour entirely eliminated.
7. A separate oil system installed for the turbo-superchargers.
8. All oxygen bottles removed from the command deck to permit the accommodation of special bombing equipment.

In addition, all H2X lead aircraft (which were initially modified in the US) were re-worked to allow the operator to face forward behind the pilot, to provide complete DR navigational facilities for the operator, adequate table space for maps and equipment, additional instruments, and a remote scope camera unit directly behind the operator which could be adjusted during the approach to the bomb run.

One final Liberator modification should be mentioned, if only to typify the many weird, wonderful and little-known indignities to which the B-24 was subjected during its wartime career. It seems that the 392nd Bomb Group, shortly after the first Me-262's appeared over Germany, decided some sort of retaliatory weapon was called for. Borrowing a truckload of bazookas from a nearby Army installation, the 392nd armament team went into action. Two of the 'stovepipes' were mounted on each of three lead and three slot aircraft, one on each side and slightly below the base of the tail turret. The weapons pointed aft and were fired electrically by the tail gunner by means of a toggle switch held in his lap. Fuses were set to detonate at 450 yd. About three days later the 392nd fired its first bazooka shots in anger. Without, as it turned out, much effect, but at least the men felt that somebody was doing *something* about Goering's latest threat to completing a tour in the European theatre.

Transport and Cargo Liberators

Faced with a war in which the enemies to be

conquered included distance and time, and for which battle no adequate preparations had been made, the USAAF was fortunate indeed to have within its arsenal of weapons one that could be readily adapted to meet the crisis. The Liberator's qualities of long range, roomy (by 1942 standards) fuselage and easy loading characteristics were quickly recognised. As usual, the problem was quantity. While some of the LB-30's and early B-24's had been modified to enhance their passenger and cargo capabilities, these changes often varied from one aircraft to the next. This was creating a highly individualised collection of transport aircraft instead of the standardised, interchangeable fleet required for planning and scheduling purposes.

The answer began in the Arizona desert, where a B-24D had crash-landed early in 1942. This aircraft was given makeshift repairs and flown, gear down, to San Diego where Liberator designer I. M. Laddon supervised its transformation to the C-87 prototype. As Laddon would later recall, 'There wasn't time for drawings. Col Brandt, the Air Force representative, and I did it mostly by waving our arms and pointing to show where we wanted equipment taken out, a deck laid or an opening cut.'

As it worked out, the B-24/C-87 conversion included the removal of all armament and bombing equipment. An access door was installed in the nose, of approximately the same shape as the greenhouse it replaced. This was hinged on the starboard side. The navigator's station was moved to the rear of the flight deck and the astrodome relocated in the centre of the faired-over section formerly occupied by the top turret. Semi-permanent flooring was installed over the bomb bays and seven passenger windows provided on each side. These were kept a uniform height from the floor, which accounted for their non-linear configuration towards the rear of the ship. Each window was holed in the centre and fitted with an adjustable device to provide a somewhat primitive means of individual ventilation control. (It has also been reported, somewhat facetiously, that these ports allowed passengers to fire their side arms at attacking fighters—which didn't bother the fighters but made the passengers feel better.) A 6 ft by 6 ft, double-doored cargo hatch was cut into the port side of the rear fuselage and the tail turret position was faired in, with a large Plexiglas area left for observation.

The prototype was flown to Bolling Field and shown to General Arnold, who immediately ordered the transport produced in quantity. Production of the new 'Liberator Express' was assigned to Fort Worth, which was just beginning to assemble B-24D's from sub-assemblies shipped from San Diego. The switch was made starting with 41-11639 and included thirty-three more serials from Contract AC-16005. When these were completed, production at Fort Worth was kept rolling without a break by sending thirty-eight more sets of sub-assemblies from San Diego, produced there under Contract DA-4. Of these, thirty-five became C-87's and three became the first C-87A's, a ten-berth sleeper version with deluxe interior appointments and slightly different window arrangement. One of these C-87A's, 41-24159, which was accepted by the USAAF on 2 June 1943, became the Presidential aircraft *Guess Where II.* (The three USAAF C-87A's had originally been known as *Gulliver's* I, II and III.)

By this time the Fort Worth plant was in series production itself, so the next C-87 consignment from San Diego consisted of twenty-seven serial numbers only. However, *new* serials were assigned at Fort Worth (42-107249/107275) and as a result no aircraft were built bearing serials 41-24312/24338.

The balance of C-87 factory production was made up of seventy-three C-87's and three more C-87A's with fiscal year 1943 serials (43-), plus a final order of 111 C-87's bearing fiscal year 1944 serials. These were cleaned up aerodynamically, particularly at the tail where a 2½ ft, cone-shaped extension was added. Twenty-four of the last 111 were allocated to the UK where they became Liberator C.VII's EW611/634. These were US serials 44-39219/226, 236, 237 and 248/261.

Because of its makeshift origin and the pressing need to produce it in quantity, the C-87 incorporated many compromises which

Above: AM927, a Liberator I, crashed in the US shortly after delivery. It was rebuilt to approximate C-87 configuration and utilized by Consolidated as a company aircraft making frequent trips between the firm's far-flung wartime enterprises. / Ken Sumney

Right: C-87 interior—not exactly first class by today's standards but better than buckets. / General Dynamics

Top: The C-87 prototype arrives at Bolling Field. / USAF

Above: C-87 number 41-11706, the thirteenth C-87 delivered from Fort Worth. / Consolidated

Above right: B-24 components being used to assemble the first three C-87A's on Fort Worth assembly line. / Consolidated

Right: One of the 25 C-87's supplied to the UK that become EW611-634. This aircraft is EW615, formerly 44-39223. Note where USAAF insignia has been painted out on rear fuselage, and jagged hole through fabric of right rudder. / USAF

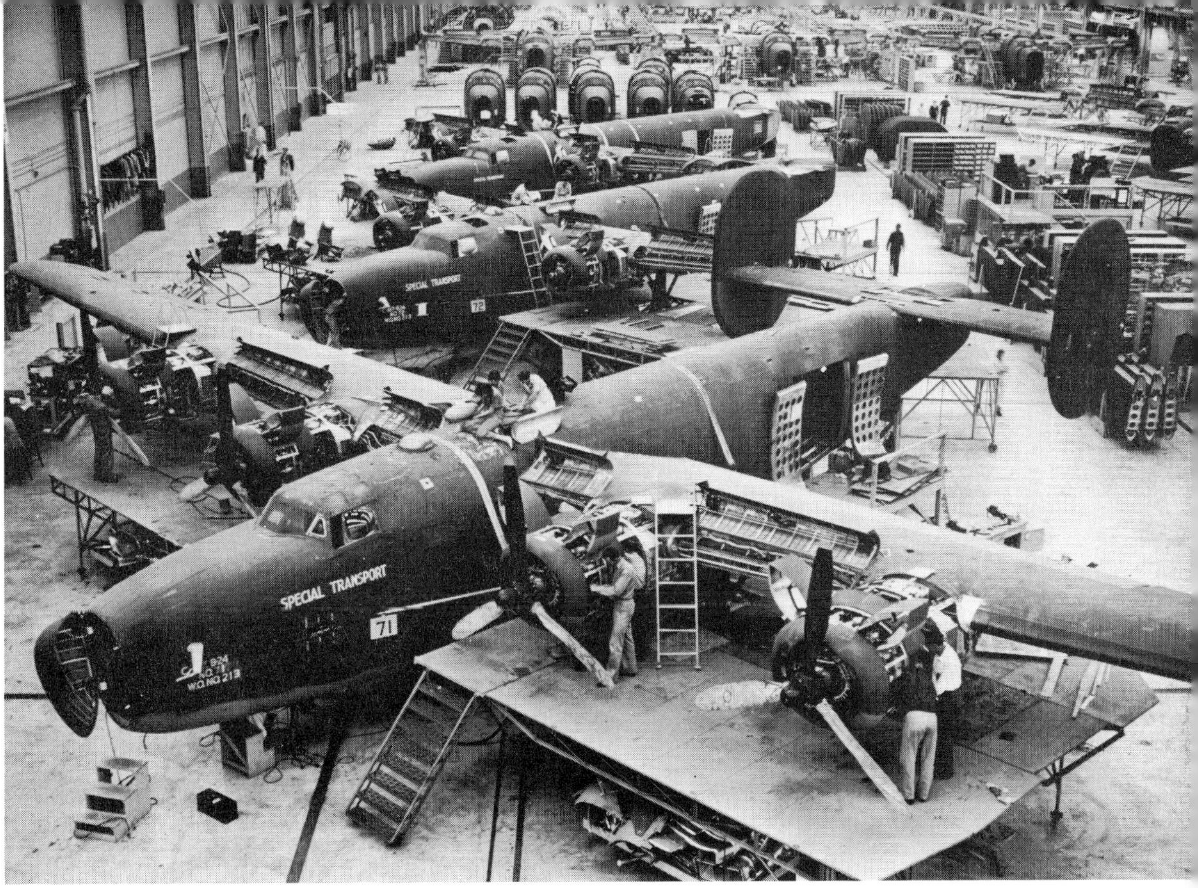
SPECIAL TRANSPORT
71
SPECIAL TRANSPORT
72

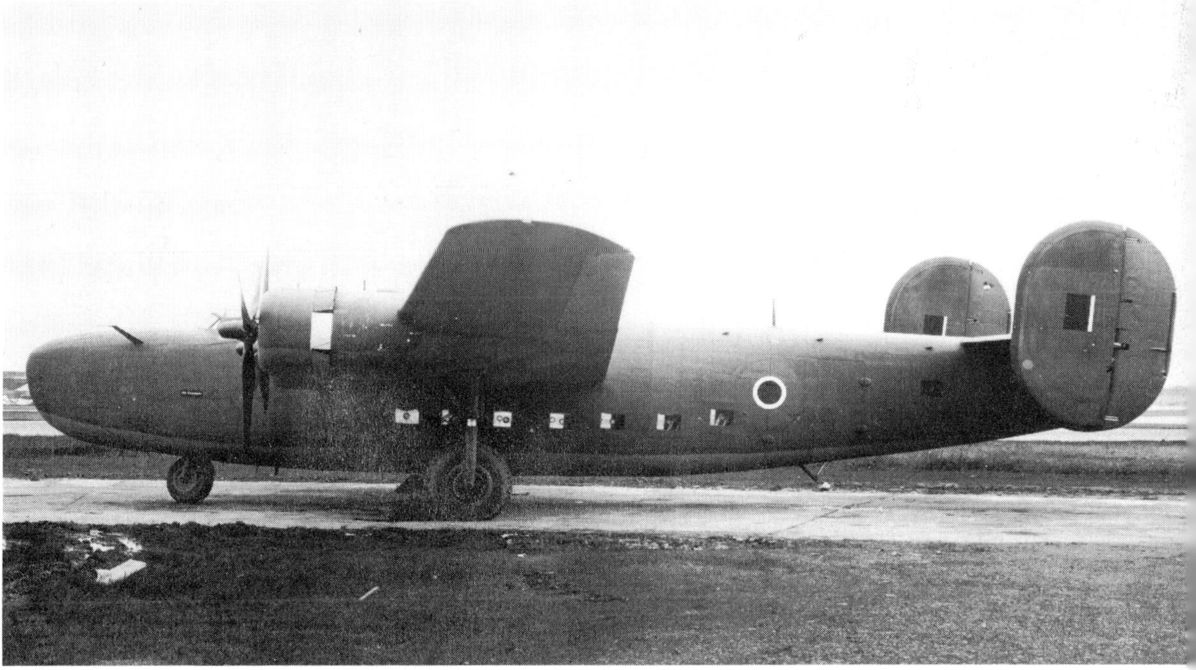

reduced its potential as a transport aircraft. It was, for example, equipped with the turbo-superchargers of the standard B-24 power package. These added 1,500 lb in weight and, at the normal operating altitude of the C-87, actually decreased cruising speed by 15 mph. Also unwanted were the self-sealing fuel cells, which added 2,000 lb, decreased range because of decreased fuel capacity, and agumented maintenance problems. Although unwanted, the cells were nevertheless necessary since Fort Worth found it next to impossible to engineer leak-tight integral tanks for the C-87 in the time available.

The other serious problem was the landing gear. While this Liberator component had proven remarkably free from troubles on the B-24, the bomber version seldom had to land with a full load as did the C-87. The nose gear in particular proved troublesome and early failures were many. The problem was corrected by replacing the actuating cylinder piston rod eyebolts with solid shank eyebolts. This took care of the structural deficiency, but did nothing to alleviate the shock of dropping five dozen tons of aircraft on to a runway. As long as they were around, the C-87's were hell on tyres.

The C-87B was an armed version not put into production. Preliminary specifications included two nose guns and the standard Martin upper turret, plus single tunnel and tail guns. This version of the C-87B should not be confused with *Pinocchio*—a unique B-24/C-87/RY-3 hybrid which, like the prototype C-87 itself, rose Phoenix-like from a crash near Tucson, Arizona to become, along with AM927 and AL610, the third member of CAC's wartime fleet of Company Liberators. *Pinocchio* was B-24D 42-40355. After its accident it was repaired and rebuilt—its nose growing in the process—and for a time was known unofficially as 'the XC-87B'. Still later a full RY-3 tail assembly was added.

A letter of intent was received in September 1943 for 125 Army C-87C transports which were to have the basic configuration of the PB4Y-2—i.e., fuselage stretch, single tail, non-turbo power plants and, as an added feature, the integral tanks which had had to be foregone on the C-87. These aircraft were allocated USAAF serial numbers 44-52853/52977. Subsequently, because of their similarity to the Privateer, it was decided to procure these aircraft under Navy contract (the order for 112 RY-3's previously described) and the USAAF serials were cancelled.

C-87 deliveries ended at Fort Worth in August 1944, after 286, including the six C-87A's, had been accepted. While this figure was increased by some non-factory conversions, particularly by the modification centre at Nashville, the total number of C-87's was an insignificant percentage of overall Liberator production.[26] The ratio was hardly the same with regard to relative importance, however. From the globe-circling diplomat to the wounded GI it was, figuratively and literally, a life-saver.

The USN took over three of the C-87A's (43-30569/571), which were designated RY-1's and given BuNo's 67797/799. The Navy also operated five C-87's as RY-2's. These aircraft, ex-AAF 43-39198/202, received BuNo's 39013/017.

An off-shoot of the C-87 was the AT-22, five of which were built on the Fort Worth line with the first, 42-107266, being accepted by the USAAF on 11 June 1943. It should be noted that these aircraft were built and accepted *as AT-22's,* and were not post-production conversions of C-87's. This variant was used to train flight engineers, a number of whom could be stationed at individual flight control panels installed within the otherwise barren interior. All AT-22's were later redesignated TB-24D's and, as such, four of them became the only B-24's to carry 43-serials.

From the records standpoint, one of the most interesting C-87's was the prototype itself, which was originally converted from the crashed B-24D-CO 41-11608. After serving in the Caribbean area, this aircraft was routed ro Olmsted Field at Middletown, Pennsylvania, for refurbishing during June 1944. When the job was completed, the ship was given a *new* serial number, 41-39600, and assigned to the Pacific Wing of the ATC. Now, this serial is one higher than the total fiscal year 1941 range

Above: ***Pinocchio:*** **Stage II. Note that LB-30 engine packages are now installed.** / General Dynamics

Below: ***Pinocchio:*** **final configuration.** / General Dynamics

Above: Last of the final USAAF order for ten C-87 aircraft, 44-52987 was delivered on 12 August 1944. Coming long after quantity C-87 production had been phased out, these ten aircraft, like the first C-87's, were converted from bomber (in this case B-24J) components on the Fort Worth assembly line. Note waist window openings have been retained. / Consolidated

Below: C-87 44-39247 at Myitkyina, Burma in December 1944. / Ken Sumney

Top right: AT-22 interior. / USAF

Centre right: As noted in the text, this is the only Liberator known to have carried two different USAAF serials, 41-11608 and 41-39600. After completing her wartime service, the original C-87—with her original serial number restored—was mothballed as shown right for preservation by the US Air Force Museum along with several other Liberators of various marks. Apparently all were later destroyed except for the B-24D currently displayed there. / USAF

Bottom right: Hoopoe, A B-24D-CF, is shown here before Eight Air Force conversions to C-87. / USAF

212

263779

for *all* US aircraft, and the reason for its assignment and use in 1944 remains a mystery. Perhaps USAAF records at Middletown indicated that 41-11608 was supposed to be a pile of B-24D wreckage in Arizona!

Besides the LB-30 conversions, which will be described later, the only 'official' B-24 to C-87 conversions (i.e., after which the individual aircraft records were changed to reflect the new designation), were: D-CO numbers 41-11680, 42-40355, 499, 503 and 552; D-CF numbers 42-63779, 780, 783, 785 and 790; and E-FO numbers 42-6976 and 985. Number 42-6976 merits special mention, for it represented Ford's own ideas of how a passenger/cargo Liberator should be configured. Dubbed the Ford Utility Transport, it lost out to the Consolidated version because by the time it was delivered on 12 November 1942, the Laddon/Brandt version was already in production at Fort Worth. This same aircraft was unusual in another way, inasmuch as it began life as a collection of spare parts. Contract ac 13281-5 with San Diego called for sixty B-24D's, with four of these to be delivered as equivalent spares. In the course of events one set of spares was shipped to Ford at Willow Run as an 'educational' aircraft, and it was this same set of B-24D parts that was eventually delivered to the USAAF by Ford as B-24E 42-6976.

The peak USAAF inventory of C-87's was reached in July 1944, when 208 were on the rolls.

In the later stages of the war, numerous Liberators were relieved from combat duty and converted for use as cargo and/or general purpose transports. These aircraft were officially redesignated as CB-24's. Some were extensively reworked and often were stripped of paint (as were most of the LB-30/C-87 conversions) to improve both their speed and appearance. An interesting twin example of this conversion involved B-24D's 41-23838 and 41-24168, the face-lifted versions of which were known far and wide as *Pacific Scamp* and *Pacific Vamp*. Other examples of CB-24's:

41-29408
42-7616 42-63787
42-40543 42-64175 (ex F-7A)
42-40939 42-73037
42-63780 42-109946
42-63781 44-40633
42-63783[27] 44-40678

The USN made an abortive effort to establish an independent source of transport aircraft with the R2Y-1. This Liberator derivative was based on the CAC Model 39, which was designed and prototyped as a completely private venture by Consolidated. Basically it was a standard PB4Y-2 fitted with a new fuselage in order to eliminate the inherent compromises involved in converting bombers to cargo carriers. The new fuselage design was of circular cross-section, 10 ft 6 in. in diameter and 90 ft long, with built-in cargo handling gear and generous capacity. Work on the Model 39 was nearing completion (the fuselage had been fabricated in Fort Worth and shipped to San Diego for the addition of its Privateer components) when the Navy, attracted by the promising idea of a relatively high-capacity cargo aircraft based largely on existing components and therefore available on a short schedule, issued a letter of intent to CVAC on 20 March 1944 for 253 of the aircraft to be designated R2Y-1. BuNo's 90132/90384 were assigned. The first aircraft was to be delivered for flight during August 1944. Two more were to follow in September, with the total contract to be completed upon delivery of fifty aircraft each during May and June 1945. Note that the Model 39 was not a part of this contract, nor was any prototype ordered. The contract was for 253 R2Y-1 aircraft. Period.

CVAC quickly completed a mock-up of the R2Y-1 and it was inspected by a USN team during the period 10-12 April 1944. After careful consideration of the resulting report the Navy rejected the aircraft based on the following major criticisms: (1) the carry-through wing structure intruded into the cargo compartment; (2) the limited landing weight and limited gross weight were disappointing; (3) the estimated length of take-off run was a serious deterrent to the aircraft's military value, and (4) the use of a wing and landing gear originally designed for an aircraft of lower gross weight appeared to limit the future development of the aircraft by the typical method of installing more powerful, but heavier, engines. Consequently, all work under Contract Noa(s) 3237 was ordered to be stopped on 6 July 1944. The assigned BuNo's were cancelled and part of the series, 90132/90271, reassigned to new PB4Y-1s.

The first R2Y-1 was not complete at the time of contract cancellation. During termination negotiations CVAC offered to buy this aircraft from the USN in its uncompleted condition. This was accepted, and CVAC paid $120,400 for the aeroplane, and an additional $93,000 for two sets of Government Furnished Equipment (GFE) installed in this aircraft and in the Model 39. This sale took place on or about 11 September 1944.

As noted, the Model 39 was built at CVAC expense. As it approached completion in April 1944, the company asked the USN to assign a serial number to this aircraft so that it could be flown prior to CVAC's obtaining certification of the aircraft from the CAA. CVAC also requested permission to fly the aircraft in military markings so that 'test flights will not be hampered in coastal defense area'. After some deliberation, the USN granted these requests on the basis that development of the R2Y-1 would be expedited and, in addition, there appeared to be 'nothing in Navy Regulations that prohibits it'.

Thus on 13 April 1944 USN BuNo 09803 was assigned to the Model 39, which was entered on the Navy serial list as an 'XR2Y-1'—even though the USN did not own the aeroplane, did not intend to buy it, or had never issued a contract for an 'XR2Y-1'. The first flight of this aricraft, which was powered by R-1830-94 engines on loan from Pratt & Whitney, took place two days later on 15 April 1944.

More about the Model 39 later on.

The F-7 Series

In December 1943 four Liberators appeared in the Pacific to continue the unfinished task that two of their ancestors, both B-24A's, had attempted just two years before. That mission, terminated prematurely by the outbreak of war, was to photograph secretly certain Japanese bastions in the Pacific. But while the end objectives remained unchanged the means had changed considerably, for the 1943 aircraft were the first of a new Liberator variant—the F-7.

Based on the prototype XF-7, which had been converted from B-24D 41-11653 at Lowry AAB in January 1943, these first four F-7's (42-40433, 476, 488 and 494) were converted by Lockheed at Dallas the following July and carried a total of eleven cameras in nose, bomb bay and rear fuselage positions. Later the same year eleven similar modifications, designated F-7A, were made at the Northwest Airlines Modification Center at St Paul, Minnesota. During 1944 Northwest modified a further 107, Martin (Omaha) eighteen, and Consolidated (Tucson) one aircraft. Of these, eighty-nine became F-7A's (three nose cameras, three bomb bay cameras) and the rest F-7B's (six bomb bay cameras). Production ended in 1945 after four special F-7B/H_2X conversions by Northwest and thirty-six standard F-7B's by Consolidated/Tucson had been completed. Total F-7 conversions were 214, including thirty-two F-7B's accomplished overseas. F-7A serials included sixty-five B-24J-CF's within the 42-64047/64371 range and twenty-four more B-24J-CO's between 42-73020 and 42-73157. The first nine F-7B's bore J-CF serials between 42-64239 and 42-64262, with all but two of the balance of 113 chosen from the B-24J-CO range beginning with 44-40147 and ending with 44-42695. The final two were ex-B-24M-FO 44-51518 and 44-51655. (A single B-24E, 42-64404, was temporarily converted to an F-7A in 1943 but was soon reconverted to its original configuration and is not included in the totals given here.) Although one F-7 appeared in the Mediterranean theatre during March and April 1944, their main use was in the Pacific where maximum inventory reached 119 in July 1945.

The typical F-7B conversion stripped the Liberator of all bombing equipment including the bomb racks. Fuel tanks were installed in the forward bomb bay to provide increased

Top: The 'XR2Y-1'. / Consolidated

Above: An F-7B, ex-B-24M-35-CO 44-42452.

With two exceptions, all F-7 series conversions were made to San Diego-produced machines. / Via Roger Besecker

range, and the aft bomb bay was riveted shut. The crawl deck was then raised to allow working clearance for attending the cameras, and multiple window apertures were cut and fitted with special protective glass. An upholstered cabin provided what comfort was possible during the typical long-duration missions on which the F-7 was employed.

The C-109

On 8 September 1943 Air Materiel Command (AMC) ordered the conversion of one B-24E-20-FO (42-7221) to 'an Aerial Tanker XC-109'. This was accomplished by stripping the aircraft of all armament, armour, and interior fittings not absolutely essential to the task ahead and installing eight specially shaped auxilliary gasoline tanks as follows: one in the nose compartment, two in each bomb bay, and three on the aft deck just behind the wing. These gave the aircraft a cargo load of 2,900 US gallons, which could be pumped out at destination in approximately one hour. Internal pumping arrangements were such that the aircraft could burn a large portion of its own cargo, which leads one to wonder just what the range of the XC-109 might actually have been with its total usable load of 4,850 gallons!

The XC-109 having been judged superior to the XC-108B, a B-17F (42-30190) converted at the same time for comparison purposes, a large-scale conversion programme was undertaken to support the fuel-hungry B-29's that were scheduled for operations against Japan from bases beyond the Hump in China. These 'production' conversions used a solid nose, whereas the XC-109—which was lost over the Hump on 31 August 1944—retained the normal greenhouse. Besides a few B-24E's, both J's and L's were used in the C-109 programme, as follows: two J-5-DT; seven J-10-DT; twenty-nine J-10-FO; thirty-six J-15-FO; eighteen J-20-FO; fifty-six L-1-FO; forty-seven L-5-FO; and thirteen L-10-FO. The total C-109 programme thus amounted to slightly more than 200 aircraft. It is interesting to note, however, that the original plan called for ten B-29 groups to be operational from China by October 1944, supported by *2,000* C-109's, with both figures to be *doubled* by May 1945.

As early as 1940 the Liberator had been envisioned as a tanker aircraft. Some testing was conducted at Eglin Field in 1941, using British equipment installed on a B-24 and a B-17, but it was felt that the method of 'hooking up'—which involved firing a line from the B-17 receiver to engage another line trailed by the B-24 tanker—left a great deal to be desired. Another interesting modification in the fuel carrying line, which did not progress beyond the planning stage, was to equip a B-24J as an aerial tanker to perform in-flight refuelling tests with a suitably modified B-29. Although the operation was considered feasible, it was not recommended because of the extensive changes that would be required in the Superfortress, including fuel transfer piping on the *exterior* surface of the wing. Another possibility examined was to have a Liberator B-24D 'Tug' tow a fuel-loaded XCG-17 glider which would then be cast loose at the point of delivery.

As it turned out, the most successful in-flight refuelling operation involving a Liberator would wait until the end of the war when, in late 1947, the British Ministry of Civil Aviation sponsored a series of tests over the North Atlantic. In this case, however, the Liberator was on the receiving end in the form of a Liberator II—this type being operated by BOAC at this time on a weekly mail and freight service between London and Montreal. Using modified Lancasters as tankers, the tests proved eminently successful but the need was soon negated by the arrival of a post-war generation of long-distance transport aircraft quite capable of trans-ocean flights without refuelling.

The Ferrets

The deadly serious game of airborne, electronic hide-and-seek that characterises today's cold war was born over three decades ago in a cold war of its own—the Aleutian Islands campaign. There, in January 1943, the first Ferret, a B-24D crammed with almost every operational radar detecting device then available, was able to chart the frequency,

Above: The 'standard' B-24 to C-109 conversion. / USAF

Below: A B-24M-15-CO being used to haul gasoline (note bomb bay tanks) over the Hump. Photographed at Myitkyina, Burma in 1944. / Ken Sumney

Right: Flight deck controls, C-109 (ex-B-24J-15-FO 42-51982). In typical automotive style, Ford Liberators carried the Ford Motor Company trademark in the centre of the control column. Inserts were removed for this photo. / USAF

Below right: Radar Counter Measure aircraft of 491st Bomb Group shows four 'fishhook' antennas, one under nose turret and three in transparent housings under fuselage. RCM ships also had multiple wire antennas running between aft fuselage and lower leading edge of each vertical tail fin. Aircraft shown also has glide path equipment installed, as indicated by post antenna just to the rear of the astrodome. / Via Mike Bailey

PUSS'N BOOTS

power and coverage of a previously-suspected Japanese radar installation on Kiska Island. Based on the successes of this and other early missions, the Ferret, or radar reconnaissance aircraft, became a USAAF fixture. Never large in actual numbers, the type played a role of extreme importance as the wizard war of electronics increased in intensity. And as it did, so the weapons themselves evolved. The B-24J Ferret was a far more sophisticated machine than the Aleutian pioneer.

Because of the small numbers involved, all Liberator Ferret conversions were hand made, with some of these accomplished overseas. In a typical J conversion, the rear bomb bay became the nerve centre of the aircraft; a padded, painted and insulated 8½ ft by 5½ ft room where two operators would spend up to nineteen hours at a stretch. The navigator, also provided with special equipment, was moved to the rear of the flight deck. Turrets were retained, and the retractable radome located just aft of the nose wheel.

The Ferrets continued their work throughout the conflict, with a number being utilised by the Twentieth Air Force in 1945 to carry out a special assignment of importance. Their task, which was completed by war's end, was to map the radars of the Nanpo Shoto chain of islands, which extended from Iwo Jima to Tokyo Bay and which were being used by the Japanese to track B-29 raids on the home islands. Total Twentieth Air Force use of the Liberator, which extended from May 1944 to August 1945, reached a maximum first-line strength of nine B-24's and twenty-six F-7's.

The Ferret was basically a passive aircraft. Its active counterpart, which first appeared in August 1943, was the Radar Counter Measures (RCM) ship. Also produced at the factories in relatively small numbers (San Diego turned out a total of 172) the success of radar jamming was such that a majority of all B-24's operating against Germany had been fitted with at least two spot-jammers by the end of the war.

TB-24's

Although the need for them is obvious, the use of first-line combat aircraft in the training role is often overlooked.

American flight training centres were located predominantly in the southern and western United States to take advantage of the excellent flying weather these locations provided. By 1944, B-24 pilots normally were receiving 105 hours of four-engine flying time prior to crew assignment. The crew then trained together for an additional ninety days in a Replacement Training Unit (RTU) before overseas commitment.

Although it was a relatively complicated aircraft for its day and a large step upward from the twin-engined AT-10's and AT-17's they had flown in advanced training school, the majority of pilots found that the Liberator was not difficult to master, for it had good flying characteristics *if properly loaded and properly trimmed.* Under these conditions it had inherent directional stability, excellent longitudinal stability over a wide range of centre-of-gravity locations, no unusual stall characteristics, and sufficient reserves of power. The B-24 became a beast only when over-burdened and/or improperly trimmed.

It was strictly an instrument aeroplane, and beginners were routinely informed that 'the only reason we put windows in it is so you can see other aircraft and mountains and tell whether the sun is shining'. Controls were heavy but not abnormally so if the all-important trim settings were correctly set up. Turns were normally limited to 45° of bank because of the rapid build-up of G-forces beyond that point. The key to good landings was 'control with power', with the ship being flown on to the runway rather than being dropped on to it.

In the 1942 era of short supply, B-24 trainers were often flown well past their normal usage lifetimes and many amassed imposing totals of airframe flight time. By 1943 returning 'war-wearies' were used to augment the training fleet. Understandably, neither of these types was particularly loved by the training establishment. By 1944, however, there were enough B-24's to go around and new aircraft were assigned directly to the training role. Considering only these later series of Liberator we find that 621 TB-24J's, 194 TB-24L's and 111 TB-24M's were specifically designated for this purpose. This is more than one out of every twelve Liberators of these types produced and, if deliveries to non-USAAF users are deducted, the ratio of trainer usage within the USAAF becomes even higher.

NOTES

25 On the subject of records, CVAC test pilot Bill Martin took a B-24, with no payload but otherwise not specially prepared, to an absolute altitude of 40,250 ft. Density altitude at this point approximated 42,000 ft.

26 Other nations, of course, converted numerous B-24's to transports and Canada actually purchased ten 'C-87 Conversion Kits' from San Diego in October 1944.

27 Previously converted to a C-87.

Above: A well-used B-24E-25-DT, serial 41-28568, sits out her few remaining days after V-J Day. / Via Hal Andrews

Below: B-24J-20-FO 44-48802 after modification for use as B-32 gunnery trainer. / General Dynamics

3 March 1945

'A pre-dawn take-off and assembly. As I was flying spare I didn't join formation until crossing the coast where I took No. 5 of the low squadron. Over Holland our No. 4 started throwing oil pretty bad. It was even bubbling out the top. We went on as I thought I could get back out ok if the engine failed. For once it was almost cloudless below.

'ME-262 jet planes were reported in the area about 30 minutes before the target. The guys were on their toes. Up ahead the sky before us was peppered with black smoke which marked where flak had fired at planes before us. We came in and a huge puff of flak burst right in front of the No. 4 ship. It was the largest burst I ever saw. For the next few minutes we had a lot of very close bursts just behind. The deputy lead had a pre-release and most of the squadron dropped. My nose gunner can be thankful that he didn't drop then—we dropped with the lead ship and right on the target! Waist reported great billows of smoke and flame were below. I think at long last we finally got that damn Benzol plant.

'Also reports from the waist indicated flak holes in the plane. On our right 3 B-24's were spinning down. One pulled out and 3 chutes were seen. 262 jets were still around and our P-51's were scrambling all over the sky. Thank the Lord for P-51's. . .'

Diary of a B-24 pilot
491st Bomb Group
European Theatre of Operations

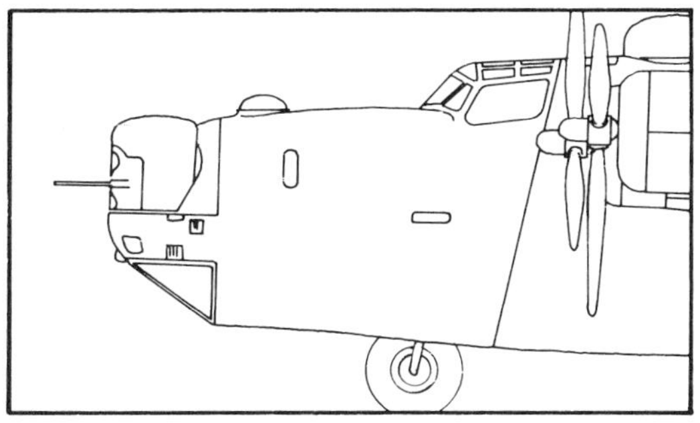

III

Operations

Early Operations

Between 10 December 1941 and 6 January 1942 the USAAF took possession of seventy-five of the Liberator II's produced for the British under Contract F-677, including FP685. All of these were carried as LB-30's (some later as C-87's) on US records and never assigned any other serial numbers. Six of these (AL527, 575, 596, 601, 607 and 621) were destroyed in US accidents within periods ranging from a few hours to a few months after delivery. Twenty-three others were returned to the UK (in lieu of an equal number of Lease Lend B-24D's scheduled for Britain)[28] between January and November 1942, with the majority changing hands in April. These aircraft were: AL507, 514, 519, 528, 529, 590, 592, 593, 595, 599, 600, 603, 614, 616, 619, 620, 624, 625, 627, 630, 635, 636 and 638. The British assigned most of these to 159, 160 and 511 Squadrons and to BOAC, where they joined other Liberator II's that had gone directly to the UK.

The remaining forty six repossessions served the USAAF in a variety of roles that in many cases spanned the entire period from the dark days of Java to victory and beyond.

Repossessed LB-30's to be used as bombers were armed by the USAAF to include a Martin turret in the mid-upper fuselage location, single magazine-fed ·50 calibre guns in tunnel, waist and nose, and twin hand-held ·50's in the tail. Non-retracting wind deflectors were placed just forward of the waist windows and additional ¼ in. armour plate installed around the tail gunner station. The British camouflage paint scheme was retained, with USAAF insignia applied.

Fifteen of these aircraft were immediately scheduled to depart for the Philippines to reinforce the beleagured 19th Bombardment Group. Although original plans called for all fifteen to go via the Atlantic-Africa-India route, only six went this way. The others left from the West Coast, stopping at Hickam, Christmas Island, Canton, Nandi-Fiji and Townsville. By the time the first aircraft of this unit (which was known as the Straubel Echelon, 7th Bomb Group) arrived, the 19th and other units of the 7th were fighting from bases in Java. Under the general code name Project X, an even dozen of these LB-30's actually arrived in Java, including all six that went via Africa, before the allied combined forces were forced to retreat to Australia. Straubel Echelon aircraft were as follows. Atlantic route: AL535, 570, 576, 608, 609 and 612. Pacific route: AL508, 515, 521, 533, 567, 572 and 606. Another aircraft, AL573, was delayed by an accident at MacDill AAB and arrived in Australia somewhat later. The fifteenth, AL607, was abandoned in flight over Wyoming before departure and its replacement, AL568, was subsequently withdrawn from the project.

The first Project X aircraft to arrive were AL609 and 612, which reached Malang on 11 January 1942. AL535 came in the next day, followed by AL576 on the 14th, AL570 on the 24th, AL508 and 521 on the 26th, and AL533 on the 30th. February arrivals saw AL515 and 567 reach Malang on the 2nd, AL608 touch down at Jogakarta on the 11th, and AL572 make Malang on the 15th. In all, seven LB-30's, including AL533, 535, 567, 572, 576, 609 and 612 are known to have been written

Top: AM920, a LIB II (note long prop hubs), converted to passenger service with BOAC as G-AHYB. De-icer boat on starboard fin has been extended to rudder, as on later Liberators, while port fin has the original version. / A. Shennan Collection

Above: Although not repossessed by the US, AL 587 did not arrive in the UK until 19 November 1942. It served with No 511 Squadron at Gibraltar and was stricken off charge on 23 March 1943. / USAF via Bob Brooks

Top: Although there undoubtedly are others, this is the only clear photograph known to the author that shows USAAF and RAF LB-30's (Liberator II's) fitted with the Martin turret amidships. Aircraft numbered 10 in background is AL596, which was later wrecked at Westover Field in January 1943. Other two LB-30's and the single B-24D are unidentified. Although this photo was taken in early 1942, note the B-17E in RAF markings in left background.

Above: A 'Project X' LB-30 enroute to Java. Note the Martin turret in the mid-upper position. Photo taken at Bangalore in early 1942. /R. A. Freeman Collection

off while based in Java while one other, AL521, was destroyed by a Japanese air raid on Darwin. AL508, 515 and 570 are known to have escaped to Australia, where they were assigned to patrol duties along with the late arrival, AL573. AL608 left Java for Ceylon on 24 February, carrying General Brereton. This aircraft subsequently participated in the first combat action of the Tenth Air Force when, on the night of 2 April 1942, two B-17's and one LB-30 of the 7th Bombardment Group were dispatched to attack enemy shipping near Port Blair in the Andaman Islands.[29]

The Far East was not the only trouble spot, and it has been little publicised that seventeen LB-30's were hurriedly fitted with Canadian-built ASV Mk II radar at Patterson Field and dispatched to Latin America to reinforce the 6th Bombardment Group whose mission was to guard the Panama Canal. Of these, seven (AL539, 543, 604, 605, 618, 623 and 631) were destroyed in operational accidents while in Latin America. Seven returned to the US in 1944; six (AL583, 628, 637, 639, 640 and 641) being converted to C-87's at Consolidated's Nashville, Tennessee facility and assigned to Consairway (see below), while AL632 was similarly assigned after a cargo-only modification by CVAC. The other Panama Patrol aircraft, AL615, 629 and 634, served out the rest of their days peacefully in Latin America where all three had been surveyed by 7 October 1944.

Alaska was threatened also, and AL602, 613 and 622 were sent north to the 28th Composite Group at Kodiak where one was wrecked on 4 June upon returning from action against the probing Japanese Fleet. In fact all three were reported destroyed by 1943, but a year later AL613 limped home and, after refurbishing, was assigned to ATC's Pacific Division with which it served as a transport until January 1946.

FP685 led a nomadic life, shuttling around half the world before being declared obsolete at Tacloban, Leyte while with the Fifth Air Force. This failed to stop it, however, and it was not finally condemned until 31 July 1946 after long and faithful transport service with Southern Cross Airways. Although in later life it bore the name *The Sad Sack* it continued, more often than not, to be referred to as *Old Number 4.*

AL589, 611, 617 and 626 were assigned to the Seventh Air Force. On 5 June 1942 they left Hickam Field for Midway under the command of Major General Clarence Tinker, Commander of the Seventh, and the next night took off to bomb Wake Island, a two-way trip of over 2,500 miles. As far as is known none of the crews found the target and AL589, with General Tinker aboard, was never seen again. Three weeks later the remaining three LB-30's tried for Wake again, this time successfully. After extensive wartime careers, mainly in transport roles after mid-1943, AL617 and 626 were written off in 1945, and AL611 broken up in July 1946.

The remaining USAAF LB-30's were pressed into service as transports. AL568 and 594 had been scheduled for return to the UK, but were re-routed to the Consolidated factory and, after suitable modification, subsequently used as original equipment on the Consairway Airline, a trans-Pacific route operated by Consolidated under contract to the Ferrying Command (later ATC). Demand for this service was so great that these aircraft, and the fifteen other Liberators that eventually joined them, were operated an average of sixteen hours per day over extended periods. AL633 was destroyed while in transport service with the Seventh Air Force. AL532, 586 and 598 were assigned to the Ferrying Command itself, in whose service all survived the war.[30]

AL610 represents somewhat of a special case. Originally scheduled for return to the British, it was detained at Wayne County Airport (a Ferrying Command stop-over point near Detroit, Michigan) and cannibalised to keep other LIB II deliveries moving on schedule. By the end of 1942 what remained of the aircraft was in such sad shape that it was considered doubtful if it could ever be made airworthy again. However, Consolidated offered to overhaul and refurbish the ship free of charge in return for the use of the aircraft as an 'inter-plant work horse' when it was not required by BAC for special duties; e.g., the ferrying of spares to units operating with the

Above: AL640, ***Jungle Queen*****, while in service with the 6th Bomb Group.** / Walter M. Jefferies Collection

Below: AL613 being refurbished for ATC service after return from Alaska. Extra length of LB-30 Curtis Electric prop hubs is well displayed in this photo. / General Dynamics

Below: FP685 as *Southern Cross Airways No.* 4. **Photo was taken as crowd gathered to marvel at the first aircraft to land on Aitutaki in the Cook Islands.** / US Army

Bottom: LB-30 FP685 in later days. / Walter M. Jefferies Collection

Right: Three AAF LB-30's. AL633 in foreground has had paint removed and is assigned to 7th Air Force. Serial of first *Southern Cross Airways* **LB-30 in background is AL626,** *Samoa*. **Serial of third aircraft,** *Australia*, **is unknown.** / USAF

Below right: AL532 after conversion to C-87. / Consolidated

LIB I or II. This proposal was approved by the Air Ministry in February 1943, and by 2 June CVAC had the completely refurbished AL610 flying again. In its new form it carried up to twenty passengers in excellent comfort, and also featured C-87 type cargo doors in the nose and port side of the fuselage. Its first Atlantic crossing, carrying both LIB II spares and passengers, took place on 7 July 1943. Similar trips followed, and the excellent impression given by these flights of AL610 were in no small measure responsible for an Air Ministry proposal in early 1944 that all LIB II's still in British service be converted to transports for use by the RAF Transport Command. The work, which was to be accomplished by Scottish Aviation, would include a fairly extensive 'upgrading' of the aircraft based in large part on CVAC's modification of AL610. A special project set up the following month under Flight Lieutenant Abbott, Engineering Officer of 511 Squadron, determined that, of the eighty-seven LIB II's[31] originally received by the RAF, forty-two were known to be serviceable (nineteen already with Transport Command, ten in the Middle East, nine in India and four with HQ Mediterranean Air Command being operated as 'Dawson Airlines') while forty-four had been struck off charge. One, AL595, could not be accounted for, its last known movement being a flight from Lydda to Heliopolis on 22 June 1943. All of the serviceable aircraft were scheduled for return and modification, together with a large number of serviceable items from the aircraft struck off charge that would be used in the process. Salvaged power plant items were particularly valuable, since these S3C4G (R-1830-61) units were, as the USAAF also had discovered with its equivalent LB-30's, essentially impossible to replace.

To return to the Ferrying Command (after our AL610 digression), this organisation was created on 29 May 1941 'to move aircraft by air from factories to such terminals as may be designated' and 'to maintain such special air ferry service as may be required'. By 7 December 1941 the Command had delivered sixty Liberators from San Diego to Montreal (and five to Cairo) where RAF ferry crews took over. Although moving aircraft continued to be an important function of the Ferrying Command and its successor, the Air Transport Command, the 'special service', or transport, role was the one that would increase far beyond what was originally envisioned. Using B-24A's, a regular service to Scotland was inaugurated in the summer of 1941 (the so-called 'Arnold Line') while other B-24A's were used to pioneer little-known worldwide routes that would become major skyways within a few short months after Pearl Harbor.

The Ferrying Command took on assignments that rivalled the most imaginative fiction, such as the plan to fly ammunition from Darwin, Australia, to the embattled Philippines in early 1942. Three Liberators were taken off the Washington-Cairo run for the purpose and, although the ammunition plan never was carried out on the scale envisioned, these B-24A's did deliver badly needed supplies to MacArthur's forces (including ammunition) and remained in the Far East until all three were lost.

One of these (40-2370) was shot down by fighters and another (40-2374) destroyed on the ground during the effective Japanese raid on Broome, Australia, on 3 March 1942. The pilot of the former, Lt Edson E. Kester, was killed. The world-wide nature of Ferrying Command operations is brought out by the fact that in September, Lt Kester had been busy successfully evading a German Dornier his Liberator encountered over the Atlantic.

Other early Liberators to see action were the twenty-three B-24D's of the incredible HALPRO mission which set out from Morrison Field, Florida, in May 1942 to bomb Japan from China and instead wound up bombing Ploesti from Fayid, Egypt on 12 June 1942—the first USAAF heavy bomber attack against Europe in World War II. The remnants of this force stayed on in the Middle East and eventually were incorporated into the First Provisional Group along with some equally well-used B-17s (and AL608) that had been brought in from Java and India by General Brereton.

Back in the US, meanwhile, with enough aircraft available to begin crew training and the promise of liberal quantities of Liberators to follow, the USAAF scheduled an increasing number of heavy bombardment groups to be equipped with the B-24. The first of these to enter combat was the 98th; which joined the remnants of the HALPRO mission in the Middle East. First operational in August 1942, from bases in Palestine and later in Egypt and Libya the crews of the 98th fought to keep Axis shipping from reaching the sands of North Africa, while the crew chiefs fought to keep the sands of North Africa from reaching the carburettors of the 98th's 'desert pink' B-24D's.

The 376th Group was activated in Palestine in October 1942, taking over and augmenting what was left of HALPRO and the First Provisional Group. The 376th began operations in November and the following month a detachment of twenty-four B-24D aircraft and crews arrived from England to augment Allied air strength in North Africa for Operation *Torch.* This detachment was from the 93rd Group, which had been operational from Alconbury since 9 October 1942 when it flew the first B-24 mission for the Eighth Air Force against Lille, France. When the African detachment returned to its official station in England (now Hardwick) they found another B-24 group—the 44th—had gone operational in November from Shipdham. England welcomed no more Liberator groups until the 389th arrived in June 1943—to be immediately dispatched to North Africa together with large detachments from the 44th and, again, the 93rd. This brought all five USAAF B-24 groups that were in the theatre to bases in North Africa to prepare for the 1 August 1943 low-level strike at Ploesti.

In the Pacific, the 90th Group's 'Jolly Rogers' set up headquarters at Iron Range, Australia, in November 1942 to begin the long, island-hopping climb to the north and victory. Its Fifth Air Force companion, initially headquartered at Torrens Creek, Australia, was the 43rd Group. This unit had used some Liberators in training, later was equipped with B-17's, and then converted to B-24D's in mid-1943.

Starting in December 1942 the 307th Group, on its way to the Thirteenth Air Force and Guadalcanal, paused at Hawaii and, while temporarily attached to the Seventh Air Force, flew several missions against Wake Island. Following the route General Tinker and his four LB-30's had pioneered, the raids were staged through Midway and represented a round trip of 5,500 miles—an auspicious beginning for the outfit that later adopted the name *Long Rangers.*

The Thirteenth Air Force also included the 5th Group, which flew B-24's mixed with B-17's until September 1943 when phase-out of the latter was completed.

The one other USAAF Liberator group to operate in the Pacific with B-24D's was the 380th. This unit was assigned to the Fifth Air Force, but actually was attached to the RAAF until the last eight months of the war. After a period of training Australian crews to operate B-24's, the 380th entered combat in May 1943 and continued to operate from Australian bases for the next eighteen months.

In the China, Burma, India theatre the Liberator was represented by the 7th Group, which moved from Australia to India in March 1942 after the retreat from Java where, as previously mentioned, it had operated with a mixed force of B-17's and LB-30's. By 1943 it had been re-equipped with new B-24D's which it operated chiefly in the Burma area. China proper was host to the 308th Group, the air echelon of which moved to Kunming via North Africa in early 1943.

By November of 1942 the Liberator outnumbered the Flying Fortress in the war against the Japanese and less than a year later, in October 1943, there were only three first-line B-17's in the entire Pacific while equivalent B-24 strength stood at 612. At that point the only aircraft at war with the Japanese in greater numbers was the P-40, the strength of which stood at 699 on 31 October 1943.

In Europe it was a different story where, on the same day, Fortresses outnumbered Liberators in the USAAF inventory by 1,205 to 315. And while the B-24 total eventually would surpass that of the B-17 in Europe also, this would not be accomplished with B-24D's. In the war against Germany, in fact, the

Top: Liberator II, being ferried to Cairo for the UK in late November, 1941, is shown being refuelled from 55-gallon drums at Natal, Brazil. RAF markings on these aircraft were changed to USAAF at Bolling Field prior to departure to allow aircraft to traverse neutral countries. Out of sixteen aircraft planned for delivery via this method, five (AL 569, 577, 574, 566 and 530) departed the US—the last on 6 December 1941—before Pearl Harbor triggered US repossession of the remainder. / Owen Clarke

Above: A B-24D and RAF LIB II at Karachi, India, 1943. / Ken Sumney

Below: The 376th Group over the Adriatic—an operational mixture of green-houses and nose turrets, desert sand and olive drab. / USAF

second USAAF raid on Ploesti—appropriately code-named *Tidal Wave*—might well be considered the high water mark of the B-24D's combat career. For the wave of Liberators that reached and broke against the Rumanian stronghold on 1 August 1943 receded in such honoured confusion that it was many weeks before the groups of which it was composed had fully recovered. And when they did their ranks were no longer composed exclusively of the familiar greenhouse-nosed Liberators, but instead were increasingly spotted with the uglier, we-mean-business faces of the turreted B-24H.

Early European theatre Liberator operations had proven inconclusive, with notable successes being interspersed with notable failures. The lack of replacements for both aircraft and crews prevented the mounting of B-24 missions of adequate size and it was not until the autumn of 1943 that the Liberator really began coming into its own in the Eighth Air Force.

In the Mediterranean, initial operations were more successful and the Liberator record in that theatre was never really in doubt. The same held true for the early offensive operations of the B-24 in the Pacific, where its range alone was enough to assure its rising ascendance. Again, however, it must be emphasised that, regardless of the heroic efforts of the early Liberator pioneers in all theatres, the full weight of the Liberator groups did not fall on Germany and Japan until the latter half of 1943. When it did it was with a vengeance—so much so that any attempt herein to chronicle the Liberator-at-war after this date must be in general, rather than specific, terms.

From this point on, then, we will not attempt to re-tell the story of heavy bombardment aviation in World War II—a stirring tale, of course, but already well recounted in various histories in as much detail as the reader could wish. Instead, our emphasis will focus primarily on two aspects of this period. The first considers the war's impact on the Liberator. How did the machine from San Diego rate when evaluated under wartime operations? What was changed as a result? The second attempts to convey the magnitude of the Liberator contribution by summarising—all too briefly, it is realised—the records of the primary US aerial organisations of World War II, the USAAF Combat Groups and the USN Combat Squadrons.

The Liberator Against Germany

The Liberator's record in Europe and the Mediterranean should speak for itself. Unfortunately it does not, for its voice was then, and is now, all but drowned out by the fanfare afforded the Flying Fortress. As the elder son and therefore heir-apparent to the credit that accompanied the long-sought recognition of daylight strategic bombardment, the B-17 was the darling of the press and, through it, the public. In much the same manner the European theatre found favour and fame over the Mediterranean, undoubtedly due in no small measure to the fact that the majority of the Press Corps found the English weather, damp and unpredictable as it may have been, immeasurably preferable to the slogging mud of Italy.

But regardless of the lack of fanfare afforded it and its crews, the Liberator was original equipment for no less than nineteen USAAF groups in Europe and fifteen groups in the Mediterranean for a total of thirty-four—some seven more than were sent into European combat with B-17's. Altogether these B-24 units flew 226,775 sorties and unloaded 452,508 tons of bombs as their contribution towards victory in Europe.

The following appreciation of the Liberator in the European theatre, written in early 1945, pretty well sums up the official attitude of the Eighth Air Force towards the B-24 at that point in time.

'The original B-24 would carry a greater bomb load, 8,000 lb against 6,000 lb, than the B-17. It would carry this load farther and was faster. Upon being put into operations in the European theatre, it was found that the armament and armour of the B-24 were inadequate and in order to operate without prohibitive losses it was necessary to make emergency modifications immediately. These

Left: A 460th Bomb Group B-24 in bad trouble.

Below left: With three turnin' and one burnin', a 460th Group Lib heads for home over the Adriatic. /USAF

Above: Ploesti was rough—high *or* low. /USAF

Below: An early (ex-HALPRO) B-24D in the Mediterranean theatre taking after its flying boat ancestors. /USAF

modifications consisted, among other things, of a full nose turret which, together with the other additions, substantially increased the weight, reduced the aerodynamic characteristics and, although increasing the fire power, eventually unacceptably reduced the overall utility of the aircraft. The load carrying capacity was reduced to 5,000 lb for the long-range, high altitude operation which is 1,000 lb less than the B-17. The speed was also reduced and as a result of increased gasoline consumption the radius of action, from greater, became substantially less than the B-17. Due to the reduced speed and increased weight the flight attitude of the plane was altered and this, together with the upward projection of the nose turret, reduced pilot vision and has been the cause of frequent collisions. Due to the addition of weight aft and the consequent rearward movement of the cg the longitudinal stability of the aeroplane was adversely affected and the aeroplane became unstable. The addition of the nose turret reduced directional stability and the B-24 became harder to fly. Spinning out of the overcast is much more common than with the B-17 and it is not as steady a bombing platform. The increased weight and poor aerodynamic characteristics further reduced the service ceiling until now it is difficult to hold a good formation, with load, above 24,000 ft. The B-17 can be flown as readily under similar conditions in formation at 28,000 ft. This means the flak losses over the same territory would be substantially greater in the B-24. Perhaps the greatest handicap to bombing efficiency in this aircraft is the space restrictions for bombardier and navigator in the nose and the interference with their forward vision, resulting from the present nose turret. It must be pointed out that (most) of our mission failures are the result of poor navigation and that inaccurate navigation through specified corridors has substantially increased our flak losses. To find and destroy small targets from high altitude both the navigator and bombardier must have adequate forward vision.

'It is my studied opinion that no minor modifications will make the B-24 a satisfactory aircraft for this theatre. I have not yet flown in or seen a B-24 mounting the Emerson ball nose turret. While this aircraft is definitely superior aerodynamically to the B-24L and undoubtedly has some improvement in vision, from the pictures, and from reports of the Eighth Air Force representative in the US, it still does not appear to provide adequate forward vision for the navigator and bombardier. Furthermore, it offers no improvement in space available in the nose for personnel and for the special equipment required here for navigation and instrument bombing. We feel that it is inferior to the aeroplane with the Bell power boost chin turret developed here as a result of combat experience and sent back to the US for consideration. Design changes even greater than those offered in the N are needed. While these may result in some temporary delay in production, the conditions under which this aircraft must operate here become so critical that it is essential that the required design changes be accomplished at the earliest possible date. These basic design changes must include the following:

1. Additional space must be made available in the nose to accommodate equipment and personnel for instrument space and visual bombing at high altitude. Additional space is also needed throughout the aeroplane to permit the most efficient operation of assigned equipment and encourage the wearing of full flying clothing and safety equipment.

2. Improved visibility is needed in the nose to permit accurate visible bombing and visual navigation at high or low altitudes when weather conditions permit. Improved visibility is needed on the flight deck to aid formation flying and increase safety in conditions of low visibility.

3. Requirements for firepower must be restudied to permit achievement of the requirements for space and visibility. In the nose, firepower must be considered secondary to performance, space and visibility. In the balance of the aircraft, fire-power remains primary but must be provided while

Above: Swift hunter, slow quarry. An Me-262 after a Liberator. /USAF

Below: Their mission nearly accomplished, 8th AF B-24's could sometimes bomb from less than 10,000 feet by 1945. /USAF

meeting the requirements for performance. *4.* In addition, improved performance is needed. We need improved stability, both longitudinally by cg control, and directionally by single tail and an aerodynamically smooth nose. We need improved rate of climb and take-off characteristics, higher ceiling, wider speed range to hold formation, greater radius of action, increased load carrying capability, and a reduction in the ratio of power required to power available under normal operating conditions to permit better three-engine performance in emergencies.'

Quite a shopping list! And, of course, one that was never to be filled to the full satisfaction of the Eighth Air Force high command. Fortunately the crews at Shipdham and Rackheath and Horsham St Faith and the many other bases that were home to the B-24 combat groups were not aware that they were fighting the war with such an inadequate weapon and so went right on winning it.

Mediterranean theatre opinion of the Liberator was generally more favourable. The early successes of HALPRO's B-24Ds and the LIB II's of RAF No. 160 Squadron while operating in support of the desperate North African fighting in the late spring and early summer of 1942 got things off to a good start, and the subsequent arrival of the aggressive 98th Group did nothing to hurt the image. The respect for the Liberator thus instilled throughout 9th Bomber Command was passed on to the Twelfth Air Force and, ultimately, to the Fifteenth.

There follows a thumbnail history of the thirty-four USAAF B-24 groups which saw operations in the European and Mediterranean theatres. The section concludes with a summary of US Navy operations in the Atlantic area with the PB4Y-1.

EUROPEAN THEATRE COMBAT GROUPS

34th Bomb Group

After serving as a training unit in the US until the end of 1943, the 34th Group moved to England in April 1944 and flew its first mission from its base at Mendlesham, England, on 23 May. The Group was one of the five in the Eighth Air Force that switched to B-17's, and the Group's last Liberator mission was on 24 August 1944. While using Liberators, the 34th dropped 3,659 tons of bombs, flew 2,018 individual sorties and lost fifteen aircraft in the course of its sixty-two missions. No claims for enemy aircraft were credited to the Group during this time.

44th Bomb Group

The 44th, flying from Shipdham, England, and from several bases in North Africa, flew more missions (343), more aircraft sorties (9,057) and dropped more tons of bombs (18,980) than any other European theatre B-24 group except the 93rd. Together these two units pioneered the use of the Liberator in Europe, with the 44th's *Eightballs* seemingly getting the worst of it for many months. Both operational losses and claims were high at 192 and 330 destroyed, 74 probables and 69 damaged, respectively. The 44th's heroic performance during the Ploesti raid of 1 August 1943 typified its dogged pursuit of duty no matter what the odds.

93rd Bomb Group

After flying antisub missions off the southern coast of the US during May-June 1942, the *Traveling Circus* moved to Hardwick, England, in the late summer of 1942—the first of the long line of USAAF Liberator groups to nest in that country during World War II. Entering combat on 9 October 1942, the 93rd flew a grand total of 396 missions, including 41 flown during three sojourns to Africa. The Group was one of the five that took the low road to Ploesti on 1 August 1943. In all, the 93rd dropped 19,004 tons of bombs, flew 9,321 individual sorties, lost 140 B-24's as a result of operational missions, and was credited with enemy aircraft claims of 93 destroyed, 41 probables and 44 damaged.

389th Bomb Group

Shortly after its arrival in England in June

Above: Liberators of the 34th Bomb Group. /USAF

Below: A B-24H-1 of the 44th Bomb Group. /USAF

Top: A 93rd Bomb Group Liberator. /Via Glen Tessmer

Above: Ground crew repair a B-24D engine of the 389th Bomb Group. /USAF

1943, the 389th was detailed to Africa for a series of raids which began on 9 July and included the low-level attack on Ploesti on 1 August. Returning to Hethel, England, the Group flew its first European mission and then sent another large detachment back to Africa during September and October. All told, the 389th flew 321 missions (including 14 from Africa) comprising 8,683 sorties, during which it dropped 17,548 tons of bombs, lost 153 Liberators and was given credit for 209 e/a destroyed, 31 probables and 45 damaged.

392nd Bomb Group
Rendle's Raiders moved to England during mid-summer of 1943, taking with them the first B-24H's off the Ford assembly line. Operations began from Wendling on 9 September 1943 and eventually reached the following totals: 285 missions, 8,015 individual sorties, and 17,452 tons dropped. Operational losses, at 184, were second only to the 44th. The 392nd was a leader, also—during the period February-May 1944 it bombed with greater accuracy than any other European theatre Liberator group. Enemy fighter claims were 144—45—49.

445th Bomb Group
The Group moved to England in late 1943 and began operations from Tibenham in mid-December. From that time until VE Day the 445th flew 282 missions, losing 133 B-24's in the course of the 8,085 individual sorties flown. Confirmed claims were 89—31—37. Total bomb tonnage dropped was 16,732. During the period June-December 1944 no other European theatre Liberator group exceeded the 445th in bombing accuracy.

446th Bomb Group
The 446th flew from Bungay, England. Entering operations in mid-December of 1944, the Group dispatched 8,180 Liberators in the course of 273 missions. It contributed 16,819 tons of bombs to the fall of Germany. Only 86 446th Liberators were lost, a number far lower than its sister groups, the 445th and 448th. Claims were also relatively low at 34—11—8.

448th Bomb Group
The third of the three Eighth Air Force B-24 units to go operational in December 1943, the 448th mounted 262 missions from its base at Seething. Before the end of the European war the Group had flown 7,707 sorties and 15,272 tons of bombs had fallen on Europe from its bomb bays. The price the 448th paid was not light, for 135 of its Liberators were destroyed in operational losses. In return, the 448th billed the Luftwaffe in the amounts of 44—19—30.

453rd Bomb Group
The 453rd took up residence at Old Buckenham, England, in December 1943 and went operational on 5 February 1944, shortly before the 'Big Week'. In the course of its 259 missions the Group flew 7,431 sorties and dropped 15,804 tons of bombs. Operational losses were 83, while 453rd gunners were awarded claims of 42—12—19.

458th Bomb Group
Horsham St Faith, England, was the home of the 458th from January 1944 until June 1945. From its runways the Group flew 240 missions consisting of 6,592 credit sorties. Tonnage dropped was 13,204, operational losses 65, and claims 28—3—14. During the period April-June 1944, the 458th led all other Liberator groups in bombing accuracy. The Group was chosen to fly experimental missions employing the Azon guided bomb during the summer of 1944.

466th Bomb Group
The 466th entered combat with a raid to Berlin on 22 March 1944. Some 231 additional missions were mounted from the Group's base at Attlebridge, England, before VE Day. The 466th contribution to that event included 6,478 bombing sorties, 12,914 tons of bombs dropped and confirmed claims of 29—3—14. Operational losses were 72 Liberators.

467th Bomb Group
The European air offensive felt the added

Top: A damaged B-24H-1 of the 392nd Bomb Group. /USAF

Above: A B-24H of the 445th Bomb Group after a forced landing. /USAF

Top: A B-24H of the 448th Bomb Group after crash landing in September 1944. /USAF

Above: Liberators of the 466th Bomb Group release their bombs. /USAF

weight of the 467th beginning on 10 April 1944. By the end of the war that weight had climbed to 13,333 tons of bombs from 6,087 individual 467th B-24 sorties. In all the Group flew 212 missions from its base at Rackheath, losing 48 aircraft while claiming 6—5—2. The 467th finished on top of the Liberator heap, for during the overall period January-April 1945 the Group led all other theatre B-24's in bombing accuracy.

486th Bomb Group

Based at Sudbury, England, the men of the 486th flew their first mission on 7 May 1944. After 46 Liberator missions the Group exchanged its B-24's for B-17's beginning in August 1944. The Group flew 1,433 individual sorties, and dropped 2,248 tons while using B-24's, losing only four Liberators in the process. No enemy aircraft claims were confirmed for the Group during the period.

487th Bomb Group

Like the 486th, the 487th also went operational on 7 May 1944 and traded its Liberators for Flying Fortresses on 2 August 1944. During this time, it also flew 46 missions. Individual sorties totalled 1,401 and bomb tonnage 2,138 tons. Eleven Liberators failed to return to the Group's base at Lavenham, England. Confirmed claims were 0—2—1.

489th Bomb Group

The 489th, based at Halesworth, England, entered combat on 30 May 1944. On 12 November 1944 it was removed from operational status to become the first group scheduled for redeployment against Japan. Its 96 combat crews were reduced to 53, and the first of these flew their B-24's back to the US in November. Although it was originally intended to re-equip the 489th with Liberators, this was changed to the B-29 and training with the new equipment was not completed by the war's end. During its European operations the Group flew 106 missions, 3,259 individual B-24 sorties, dropped 6,951 tons of bombs, lost 41 Liberators and was credited with a single enemy fighter destroyed.

490th Bomb Group

Based at Eye, England, beginning in May 1944, the 490th flew its first mission on the 31st of that month. Thirty-nine more followed, the last on 27 August, before the Group exchanged its B-24's for B-17's. During this time, the 490th flew 1,117 Liberator sorties, dropped 1,792 tons of bombs and lost 2 aircraft. No enemy aircraft claims were awarded.

491st Bomb Group

This Group was the last one scheduled for operations with the Eighth Air Force, although it flew its first mission on 2 June 1944, a few days ahead of the 493rd. Operating initially from Metfield, England, and later from North Pickenham the 491st flew 187 missions, losing 70 aircraft in the process. Bombs dropped totalled 12,304 tons; individual sorties totalled 5,548; confirmed claims were 9—10—3.

492nd Bomb Group

The 492nd, based at North Pickenham, England, entered combat on 11 May 1944 and soon achieved the unenvied reputation of being a 'hard luck' outfit. By early August the Group had flown 64 missions, 1,606 individual sorties and dropped 3,757 tons of bombs—but had lost 57 Liberators in the process while receiving credit for claims of 21—0—3. In a general contraction of the number of B-24-equipped groups in the Eighth Air Force, the 492nd was taken off daylight bombardment and assigned to *Carpetbagger* and other special operations. These missions, which were also flown for the large part in Liberators, were carried out from a new base at Harrington.

493rd Bomb Group

The 493rd's Liberators arrived at Debach, England, in April 1944 and the Group went operational on D-Day—the last Eighth Air Force group to do so. Between that date and the end of August 1944 the Group flew 47 missions, 1,478 sorties, lost 11 Liberators and dropped 2,457 tons of bombs. The 493rd operated with B-17's thereafter.

MEDITERRANEAN THEATRE COMBAT GROUPS

98th Bomb Group
Activated in February 1942, the 98th moved to the Mediterranean theatre the following summer and remained in action in that theatre throughout the war, being assigned in turn to the Ninth, Twelfth and Fifteenth Air Forces. Largest contributor to the low-level Ploesti mission, the 98th continued to serve with distinction over a variety of targets that included many of the toughest on the Continent. The 98th flew a total of 417 missions.

376th Bomb Group
The *Liberandos* were activated in Palestine in October 1942. Like the 98th, the 376th served with distinction in the Ninth, Twelfth and Fifteenth Air Forces. The Group earned three Distinguished Unit Citations, one of which was for its participation in the low-level assault on Ploesti. The 376th flew more missions—451—than any other Liberator bombardment group in the European and Mediterranean theatres.

449th Bomb Group
Arriving in Italy near the end of 1943, the 449th flew its initial mission on 10 December of that year. It was the first of over 200 for the Group, including a pair (Bucharest and Ploesti) for which the 449th was awarded two Distinguished Unit Citations. The Group operated out of Grottaglie.

450th Bomb Group
In many ways the twin of the 449th, the 450th accompanied the former to Italy (Manduria) in December 1943 and also flew its first mission on the 10th of that month. Like the 449th it, too, won two DUC's for performance on two particularly rough missions over Regensburg and Ploesti. The Group was known as the *Cotton Tails* because of their original white rudder markings. Companions to the last, the 449th and 450th were re-deployed to the US together on 15 May 1945.

451st Bomb Group
The 451st began operations with the Fifteenth Air Force in January 1944 and by the end of the war had completed 216 missions, flown 5,568 effective sorties, dropped 11,961 tons of bombs and lost 135 Liberators. Three Distinguished Unit Citations were awarded the Group for missions against Regensburg, Ploesti and Vienna. Enemy claims were 98—41—13. The 451st was initially based at Gioia del Colle, later moving to San Pancrazio and finally to Castelluccia.

454th Bomb Group
Entering combat during January 1944, the 454th completed 243 missions before VE Day, with trips to Bad Voslau and Linz resulting in Distinguished Unit Citations for the Group. The aircraft illustrated crashed on take-off from the Group's base at San Giovanni.

455th Bomb Group
Based at San Giovanni, where it arrived in January 1944, the 455th flew over 200 missions in support of the destruction of the German war effort. The Group was awarded Distinguished Unit Citations for missions against Steyr in April 1944 and Moosbierbaum just two months later. For the month of January 1945 the 455th led the Fifteenth Air Force in bombing accuracy. The only group to stay on in the Mediterranean theatre after the close of hostilities, the 455th was inactivated in Italy in September 1945.

456th Bomb Group
The last of the three groups to enter operations with the Fifteenth Air Force in the early months of 1944 (the others were the 454th and 455th) the 456th was based at Stornara. Like its sister Squadrons, the Group flew in excess of 200 missions and for two of them received Distinguished Unit Citations—the first, on 10 May 1944, for the classic action of fighting through and bombing the target after other groups had turned back because of weather. The Group led the Fifteenth Air Force in bombing during March 1945. It returned to the States in July of that year.

Above: A B-24M of the 451st Bomb Group releases its bombs over the target. /USAF

Below: A wrecked B-24M of the 454th Bomb Group. /USAF

459th Bomb Group
The 459th flew its first operational mission from its base at Jiulia in March 1944. From the start it suffered heavy losses compared to the other Fifteenth Air Force groups, losing 34 Liberators (all causes) during the first three months and a startling 67 during June, July and August. Claims during this six-month period were equally high at 86—36—23. Fortunately, losses were reduced during later months and the 459th doled out more than it took as the war reached its finale. It returned to the US on July 1945.

460th Bomb Group
After training at Chatham Field, Georgia, the 460th departed for Spinazolla, Italy, in early 1944 and flew its first combat mission on 19 March. The Group was awarded a Distinguished Unit Citation for leading the 55th Bomb Wing Liberators through bad weather and flak to a successful attack on Zwolfaxing on 26 July 1944. After the close of hostilities the 460th was assigned to the Air Transport Command to assist in returning personnel from Europe to the US. It was inactivated in Brazil in September 1945.

461st Bomb Group
The Group began combat operations in April 1944. By war's end, it had flown 214 missions, dropped 10,844 tons of bombs, lost 29 Liberators to enemy aircraft, 46 to flak and 47 to other causes. Claims were 112—49—21. Based at Torretto, the 461st received two Distinguished Unit Citations. It returned to the US in June 1945.

464th Bomb Group
Beginning operations in April 1944, the 464th completed 187 missions against *Festung Europa* in the twelve months before Germany called it quits. The Group was based at Panatella and received Distinguished Unit Citations for missions against Vienna on 8 July 1944 and Pardubice on 24 August 1944. Like the 460th, the 464th was assigned to ATC services after VE Day.

465th Bomb Group
Also based at Panatella, the 465th went operational on 5 May 1944. Total missions were 191 involving 4,749 effective individual sorties and 10,528 tons-on-target. Enemy aircraft claims were 72—18—32. Two Distinguished Unit Citations were awarded to the Group, one in July and one in August 1944. The Group also carried out operations in support of the Yugoslav partisans by bombing troop concentration camps and bivouac areas in that country during its first month of combat operations. Assigned to ATC after the close of hostilities, the Group operated out of Trinidad as a part of the homecoming airlift operation.

484th Bomb Group
The *Bow Tie* Liberators began operations with the Fifteenth Air Force from Panatella in April 1944 and had completed 206 missions by the war's end. The Group flew 5,712 effective sorties, dropped 11,198 tons of bombs, lost 86 B-24's in the process and claimed 40—13—19 of the enemy. The Group was awarded Distinguished Unit Citations for missions to Innsbruck and Vienna during June and August 1944, respectively. The 484th was assigned to ATC services after the war, operating out of Casablanca. It was inactivated there on 25 July 1945.

485th Bomb Group
The 485th was based at Venosa, from which it flew its first operational mission during May 1944. The Group was awarded a Distinguished Unit Citation for outstanding performance on a mission to Vienna on 26 June 1944. It returned to the US just after VE Day and was being equipped with B-29's for service against Japan at the end of hostilities. After serving with SAC, the Group was inactivated in August 1946.

Navy Operations
In the war against the U-Boats, PB4Y-1's of Fairwing (Fleet Air Wing) 7 in Iceland and Fairwing 15 in French Morocco were in operation by April 1943. Encounters between FW 200's and the Liberators of Fairwing 15

Top: A B-24H-15-DT of the 455th Bomb Group. / Via John Preston

Above: A B-24G of the 456th Bomb Group. / USAF

Top: A B-24M of the 461st Bomb Group. / Via Stan Staples

Above: A Liberator of the 464th Bomb Group. / Via Griff Murphey

were fairly commonplace, although 'kills' by either side were not frequent. In August two PB4Y-1 squadrons of Fairwing 7, VB-103 and VB-105, moved to St Eval, England, and by the end of September had flown 1,351 hours of antisub patrols over the Bay of Biscay, losing 2 of their number to JU88's in the process. Some of these Liberators were fitted with a US equivalent of the British Leigh Light for night operations.

In a less active—although hardly less strategic—area, VB-107 Liberators of Fairwing 16 covered the South Atlantic from Ascension Island beginning on 30 September 1943. Together with Liberators operated in the Northern Atlantic, including those of Canada and Britain, the Liberator now ringed the Atlantic and, when bases were later secured on San Miguel Island in the Azores, PB4Y-1's moved in at once to provide the centre of an umbrella that covered the entire hunting ground of the U-Boat.

Besides the squadrons already mentioned VB-110, VB-111, VB-112, VB-113 and VB-114—all of which received their first Liberators in October-November 1943—served on Atlantic operations. VB-125 and VB-163 were assigned PB4Y-1's in February 1944 and December 1943 respectively, but retained these aircraft for only brief periods.

The Liberator Against Japan

The B-17 versus B-24 controversy existed only where the two aircraft were flown side by side. When Liberators were operated exclusively, one was hard pressed to find any significant objection to the B-24 as a combat aircraft. And, after the opening phases of the war, such was the case in the Pacific. There, until the arrival of the B-29's late in the war, the Liberator reigned—if not altogether supreme then certainly without peer.

As is often pointed out, it was a different kind of war. The fighters weren't as numerous and the flak wasn't as thick—but then thirty aircraft over a target was a good-sized mission and it may well be that on occasion the per capita distribution of each was as high as in the European theatre. Then there was all that water. If the English Channel and the Mediterranean were enough to make the European flyer nervous, then the vast reaches of the Pacific should have been more than the average crew could take. Of course they were not—and the crews soon grew accustomed to the fact that, as long as you're airborne, what passes under the wings is of no consequence.

It should not be inferred from the foregoing that the Lib entered the Pacific war as a perfected weapon, for such was not the case. Basically sound, yes, but lacking many of the refinements that represented the difference between the drawing board and the real world of war. At the risk of overemphasising these shortcomings, there follows the text of an early 1943 evaluation of the B-24 in the Pacific theatre. Note the emphasis on increased armament and the inevitable difference of opinions between the USAAF and USN.

Fiji. Personnel in the Second Island Command, Fiji, operate the B-24 on long-range reconnaissance missions and on strike missions out of Guadalcanal. The aircraft is looked upon with favour because of its large bomb carrying capacity and long range, although combat losses for this type are higher than for the B-17. It is felt that the nose turret and the retractable ball turret will help this latter situation since the Japanese attacks recently have been almost entirely from head-on or below. There is a general feeling that the B-24 will not absorb as much enemy fire as the B-17 although no concrete evidence could be found to support this view. The present hand-held front guns have proven entirely inadequate against Zero attacks and the bottom scarf mount, which has been installed locally, is not entirely satisfactory either because of its limited cone of fire and vision.

Brisbane. The opinion is that the B-24 is at present operationally unsatisfactory because of inadequate fire power forward and downward. It is felt that if ·50 calibre guns can be installed in the nose and tail which can be depressed at large angles, this arrangement, plus a scarf mounted belly gun, should be satisfactory. The application of armour to protect the crew

Above: 465th Bomb Group Liberator. / USAF

Below: A Liberator of the 485th Bomb Group leaves its target. / USAF

Above: A PB4Y-1 over England enroute to the Bay of Biscay for antisubmarine patrol, November 1943. / USN

Below: Leigh Light, of five million candlepower, was installed beneath Liberator starboard wing to aid in night sub-hunting. / British Air Ministry

is considered helpful, but not a substitute for increased fire power. The nose turret will improve the aircraft, both as a tactical weapon and as a vehicle, since the cg will be moved forward thus correcting a troublesome condition both in flight and on the ground.

Port Moresby. Major General Whitehead considers the B-24 a good tactical aircraft although it requires a better aerodrome than the B-17. As a result, from poor strips the B-24 carries only 4,000 lb of bomb load—or one-half of its design capacity. The relatively high rate of loss of B-24's in combat is largely due to the fact that they are used principally on solo reconnaissance and are commonly attacked by large numbers of Zeros. Personnel of the 90th Bomb Group appear to be somewhat dispirited because of their high rate of loss. These losses have been largely due to weather, and to taxying, take-offs and landing accidents. Nevertheless, they do sustain a relatively high role of combat loss in addition because of Zero attacks. Some means of protecting the crew from frontal fire is desired as is armour to protect the tail gunner's back. The latter is hit by shots aimed at forward portions of the aeroplane. The effect of frontal attack was observed on aircraft returning from the Bismark Sea engagement. Windshields were shot out and bursts through the nose were common. It was noted that a cannon burst fired head-on into the bombardier's compartment will kill everyone forward of the fuselage waist.

Eagle Farm and Archer Field. The combat weaknesses of the B-24 is under vigorous study at these locations. At Eagle Farm there are two types of nose turrets mocked up. One is very similar to the Bendix chin installation, except that the operator revolves in azimuth with his guns—a feature upon which much stress is placed. The second is a forehead type showing considerable promise. Drawings of both of these turrets are being furnished Wright Field by the 5th Air Service Command. These developments are the projects of Major Gunn, a gadgetering engineer of unusual talent.

At Archer Field a Consolidated tail turret is being installed in the nose of a B-24 by personnel of the 5th Air Service Command. This is in addition to the Consolidated turret in the tail, and a scarf mounted pair of ·50 calibre guns in the belly. The nose turret installation looks very promising. It streamlines well into the fuselage contour, gives good angles of fire, armours the bombardier and does not interfere with the acitivity of the latter. The arrangement has obvious advantages in supply, maintenance and in the training of operators. The project was undertaken at General Kenney's direction after his return from the New Guinea front where he looked into the tactical weaknesses of the aeroplane. A second B-24 was under similar modification on 10 March 1943, on which day the first was scheduled for flight test.

The B-24 tail bumper has proven unsatisfactory in service and is being replaced with a tail wheel. A Lockheed Hudson tail wheel assembly is installed on the aeroplane just aft of the main horizontal tail bulkhead. The wheel is left in the extended position. No appreciable reduction in performance has resulted from the addition of the tail wheel.

The demand for engine carriers is extremely acute in the south and south-west Pacific theatres. An LB-30 is being modified at Archer Field to take a large cargo door. No engineering computation has been done on the project. It is strictly 'cut and try'. A 6 in. by 6 in. channel is placed over the door top, but is not tied into the bulkheads nor gussetted at its extremities to the skin. It looks very bad, but no one seems concerned so long as an engine can be got into a fuselage.

Guadalcanal. The B-24's operating out of Guadalcanal carry the larger end of the burden of strike missions. Their losses have been relatively high due in part to inexperienced crews, but mostly to basic deficiencies in the aeroplane itself. Pilots feel that visibility is very poor. The possible versatility of bomb loads is also disappointing and for that reason B-17 type racks have been installed in the fuselage opposite the standard racks, permitting the aeroplane to carry up to forty 100 lb bombs. It

Left: Up, up and away. . .*Dragon Lady* of the 11th Bomb Group takes off from Guam. / USAF

Top: Scratch one B-24. / Via Tom Shutt

Above: End of the trail—a B-24D derelict photographed in Bootless Inlet near Port Moresby on 13 July 1944. / US Army

is felt that a turret in the nose of this aeroplane is an absolute necessity and an installation of a Consolidated turret in the nose essentially the same as that developed at Archer Field has been devised. This modification is being made at the Hawaiian Air Depot. Scarf mounted belly guns are used on these aircraft, but with disappointing results. It is felt that the retractable ball turret must be installed if losses are not to be excessive. In order to maintain the balance of the aeroplane, the 307th Heavy Bomb Group concurs in the idea that the tail turret may be removed and replaced by two hand-held stinger guns. This installation is also being worked out at the Hawaiian Air Depot and appears to be very successful. The B-24's at Guadalcanal have had a good deal of bomb rack trouble. All of the crews without exception have bitter comments to make about the 'five and ten cent store' quality of the bomb racks and allied equipment. All of the bomb release system brackets weave and distort at their bases, all of the linkages are spongy and tend to bend sideways at the bushings. The A-2 releases (which are made with pressed steel cases) are springy and unreliable, resulting in safe-dropping or premature dropping of the bombs. The local engineering officers have made modifications with whatever scrap material can be found. As a result there are some very peculiar bits of patching and beefing on the Guadalcanal B-24 aircraft.

It is believed necessary that bullet-proof glass be installed for the protection of the pilot, co-pilot and bombardier, but if there is any choice between armament and armour, reliance for protection should be placed in increased armament. Pilots generally of the 307th Bomb Group doubt the practicability of personal armour although several of the pilots are equipped with 3/8 in. steel shields which they rest against their chests. They desire seat armour of a form fitting type fixed to the structure of the aeroplane and not adjustable. The lack of seat adjustment will be acceptable in return for the added protection.

The B-24 has proven to have bad crash characteristics in water landings. Apparently the safest place for the crew to be is as far aft in the tail of the aeroplane as possible and it is desired that some means for belting personnel down be provided in that point to take loads both fore and aft. Missions have failed as a result of creeping of bomb bay doors. This situation was corrected by adjusting the door safety solenoid.

Espiritu Santo. A total of thirty-one changes in equipment and modifications are strongly recommended for the B-24 by the Navy and by the 13th Bomber Command. Until the major changes involving deficient armament are incorporated, it is considered that this type of aircraft is not suitable for daytime bombardment missions over areas moderately defended by aircraft. (*Note:* The B-24's assigned to the Navy are all sitting on the ground at Tontouta, New Caledonia.) The most important changes requested are:

6. Provide sight for night and low altitude bombing.
8. Install Consolidated type turret in nose of aeroplane.
9. Cut 2 in. from waist gun stands to increase depression and reinforce to reduce vibrations from firing.
12. Add retractable ball turret to aeroplane.
21. Rework the bomb release system. Failures occur solely to lost motions in mechanical linkages and lack of rigidity in entire bomb control set up. This system should be entirely reworked and the 'dime store' equipment replaced with quality material. The system should be rebushed and provided with support and bracing for all components. The Navy desires that the navigator be located on the flight deck behind the pilot, by regrouping the radio and radar equipment. Army personnel, however, object vigorously to this proposed modification.
22. Increase armour protection for nose and rear quarter attacks.
23. Add bullet-proof windshields, and side glass with tear drop scanning windshield for rear vision in sliding windows.
24. Increase capacity of all ammunition cans.

25. Remove the navigator's dome.
31. Install nose turret with two ·50 calibre turret guns. Such turret is considered an absolute necessity. According to the Navy and to the Bomber Command this is the most urgent of all recommended changes.

As we have already seen, most of these recommendations were implemented in short order, first at modification centres and then on the production lines. The results were so favourable that one group commander was able to state, only two months after the above report was written, that 'The improvements have greatly increased the respect of the combat men for the B-24, and we are all convinced that it is now a far better heavy bombardment plane than the B-17.'

From then on things got better. General George C. Kenney, one of the most successful employers of the Liberator, has summed up the overall role of the Liberator in the Pacific as follows:

'As a heavy bomber for use in the Pacific during World War II, the B-24 was the best thing we had until the B-29 was ready. The B-24 carried more bombs than the B-17. It could be flown while greatly overloaded, so it had much greater range than the B-17. We repeatedly hit targets over 1,200 miles from the take-off point. Although designed for a take-off weight of around 50,000 pounds, on our long range missions our take-off weight was over 70,000 pounds. The B-24 was well-built and would take a surprising amount of punishment and still stay together. In formation, and especially after we installed a power-operated turret with a pair of ·50 calibre guns in the nose, the airplane could give an excellent account of itself in combat with the Japanese fighters. However, like all bombers, it needed, and whenever possible we provided, fighter escort on daylight missions to keep from taking losses that in the long run might prove destructive to morale.

'As an individual airplane, it was not as nice a flying machine as the B-17. It never seemed ready to take off by itself no matter how far you let it run. You simply built up your airspeed high enough and then pulled the airplane off the ground. Its performance at high altitude was inferior to the B-17 but that made little difference to us in the Pacific where all our combat operations were either on the deck or around 6,000 feet, just above the small arms fire and too low to be a good target for the heavier guns of 77mm and up.

'The nose of the B-24 originally came with individual ·50 calibre machine guns mounted to fire through eye-ball sockets. Only one of them could be used at a time and the field of fire was quite limited. With these guns, plus a navigator and a bombardier, the nose cockpit was too crowded. When the Japs found out that we had good defensive fire power to the rear, they made their attacks head on from the front where the B-24 was weak. Accordingly, we took a tail power-operated turret from a wrecked B-24 and installed it in the nose. As soon as we tried it out in combat and found it to be highly successful, we remodelled all our B-24's in the same way as fast as we could get extra tail turrets and asked General Arnold to have all future production of B-24's in the United States fixed up the same way, with power turrets in both nose and tail. This was done and the B-24 bomber crews became quite enthusiastic about their ability to survive attacks by the Japanese fighters. The records certainly proved it.

'In order to save weight and give better take-off and flying characteristics, we took out all waist guns as our studies showed that they never hit any Jap planes anyhow and therefore, the gunner, his guns and his ammunition were simply dead weight. Some of the squadrons did away with all armor, saying that they never saw any of it with even a dent and therefore it, too, was just weight.

'All things considered, our bomber crews were quite happy with the B-24. They didn't boast about its flying qualities but they appreciated its range and load-carrying capacity. And, after all, those were the qualities that really counted in getting the job done.'

In doing that job, Pacific USAAF Liberators mounted over 87,000 sorties and dropped nearly 180,000 tons of bombs. The units that accomplished this record are briefly described and illustrated in the section that follows.

5th Bomb Group
Based in Hawaii and equipped with B-17's and B-18's at the time of Pearl Harbor, the 5th phased over to the Liberator gradually, beginning in June 1943 and completing the conversion on 25 September. The group claimed the distinction of being the first B-24 outfit to modify its aircraft to take a nose turret. Assigned to the Thirteenth Air Force, the *Bomber Barons* of the 5th followed the war northward beginning with Guadalcanal in August 1943 and continuing through New Georgia, Los Negros, Wadke, Noemfoor, Morotai and finally, Samar.

7th Bomb Group
After participating in the brief, valiant but hopeless Java campaign, the 7th moved to India where it became the heavy bombardment arm of the Tenth Air Force. By late 1942 Liberators had replaced its mixed force of well-worn B-17's and LB-30's and for the rest of the war the Group, operating from various bases in India, ranged far and wide over south-east Asia in successful pursuit of the war against Japan. The 7th was the war's most successful employer of the Azon guided bomb.

11th Bomb Group
The 11th, based at Hickam Field, Hawaii, on 7 December 1941, was in the war from the start and suffered badly during the Japanese attack. New B-17's were forthcoming and the Group flew search and patrol operations from Hawaii until mid-1942 when it was moved to the south-west Pacific. The 11th returned to Hawaii in March 1943 to be re-equipped with B-24's. Assigned to the mid-Pacific Seventh Air Force, the 11th moved westward via Funafuti, Tarawa, Kwajalein, Guam and—near the end of the war—Okinawa.

22nd Bomb Group
A medium bomber outfit until February 1944, the 22nd was one of three Liberator groups that formed the long-distance striking arm of the Fifth Air Force. Beginning B-24 operations out of Nadzab, New Guinea, the *Red Raiders* went northward via Owi, Leyte, Angaur, Samar, Luzon and Okinawa—arriving at the last named only days before the end of the war but still in time to fly a few missions over the Japanese homeland.

28th Bomb Group
Originally designed as a Composite Group, the 28th, as Alaska's Air Force, flew a variety of aircraft throughout the war. Its heavy bombardment squadron, the 404th, flew its Liberators on 2,578 sorties during which 2,812 tons of bombs were dropped and 29 enemy aircraft destroyed in the air. Losses were 33 Liberators on combat missions including 17 to enemy aircraft, 6 to flak and 10 to other causes. An additional 22 B-24's were lost to accidents. Besides the bombing role, the 28th's Liberators specialised in photo and radar reconnaissance as well as sea search and rescue. It has been little publicised that 28th Group Liberators hit Japan's home islands as early as 18 July 1943.

30th Bomb Group
After patrol duties in the US, the 30th moved to Hawaii in October 1943 and began operations from the Ellice Islands a month later. From there on it was 'one damned island after another' (as the Seventh Air Force history so aptly put it) as the 30th and its companion groups in the Seventh slugged their way towards Japan via the Central Pacific island route.

43rd Bomb Group
Although it had used some LB-30 aircraft in training, the 43rd took B-17's into its first combat. Conversion to Liberators took place during the period May-September 1943 while the Group was stationed at Dobodura, New Guinea. The Group took on the complete variety of missions that characterised the employment of the heavies in the south-west Pacific, moving northward with the war until

Above: A B-24J of the 30th Bomb Group wings its way toward the target. /USAF

Below: A Liberator of the 308th Bomb Group over its target. /USAF

Above: BuNo 65385, ***Brown Bagger's Retreat.*** **Ex B-24L-20-CO 44-41800.** / Via Hal Andrews

Below: Unusual nose turret installation on early Pacific PB4Y-1, BuNo 31987 (ex AAF and B-24D 42-40121) featured sliding section which was retracted rearwards during turret operation. Crew member on extreme left is Bob McGuire, who started the Liberator Club some 25 years after this photo was taken. / Liberator Club

its final missions, flown from Ie Shima, were against Japan itself.

90th Bomb Group
The *Jolly Rogers* entered combat in the south-west Pacific war from Iron Range, Australia, in November 1942. Like its Fifth Air Force companions, it ranged far and wide over Japanese-held territory from Balikpapan to the Phillippines. The 90th ended the war in Ie Shima alongside the 43rd.

307th Bomb Group
After training and patrol duty in the US, the 307th received new B-24's on 20 October 1942 and flew to Hickam Field, Hawaii, where it replaced the 90th Bomb Group in the Seventh Air Force. Its first mission, on 21 December 1942, was staged through Midway against Wake Island. In mid-February 1943 the Group moved south-west and was assigned to the Thirteenth Air Force based at Guadalcanal. The *Long Rangers* later moved to New Georgia in January 1944 and, after that, to a variety of bases until it reached Clark Field on Luzon in 1945.

308th Bomb Group
The only Liberator group to operate out of China, the 308th arrived at Kunming in early 1943. Supporting itself from India largely by its own supply missions over The Hump, the Group ranged over enemy territory from Saigon to Shanghai. During the last stages of the war, the 308th returned to India where it flew some joint missions with the 7th Bomb Group and also returned to ferrying gasoline and supplies into China. The Liberator shown is on a mission to Vinh, some 160 miles south of Hanoi on the Tonkin Gulf.

380th Bomb Group
The 380th Group, which called themselves *The Flying Circus*, moved overseas during April-May 1943. Initial operations were carried out from desolate Fenton and Manbulloo Fields in Northern Australia. Although assigned to the Fifth Air Force, the 380th flew under RAAF operational control until January 1945 when it moved to Mindoro, and, later, Okinawa. The Group is probably best known for its series of long-range (2,700 mile, 16 hour) attacks on Balikpapan carried out in August 1943.

494th Bomb Group
The last Liberator Group to be activated, the 494th—*Kelly's Kobras*—was a high-spirited outfit whose Liberators sported some of the outstanding nose art of the war. Assigned to the Seventh Air Force, the Group entered combat on 3 November 1944. All told, the *Kobras* flew 146 missions, 3,172 individual sorties, dropped 6,429 tons of bombs and lost 30 Liberators in the process. The Group ended the war on Okinawa, where its last strikes were directed towards China and Korea as well as Japan proper.

Navy Operations
The Navy, as the Army Air Force had done before it, assigned its first Liberators to the job of photo-reconnaissance. In this case it was a Marine Squadron, VMD-254, that got the nod. Dispatched to Espiritu Santo in mid-October 1942, VMD-254 eventually flew over 300 photo-reconnaissance sorties from Espiritu and Guadalcanal before it returned to the US in January 1944. Other Marine photo-recce units which saw action with the PB4Y-1 were VMD-154, which received its first Liberator during January 1943, and VMD-354, which got its aircraft one year later. Marine PB4Y-1's also served with MAG-14, MAG-15 and MAG-35.

In all, the USN eventually converted no less than sixty-five of its Liberators to the photographic version, or PB4Y-1P. In addition to the Marine squadrons mentioned above, the Navy equipped VD-1 (January 1943), VD-3 (February 1943), VD-4 (August 1943) and VD-5 (July 1944) with these aircraft—although VD-5 kept its Liberators only a short period. All of the aforementioned Navy and Marine squadrons operated in the Pacific.

Against the Japanese the PB4Y-1 was used extensively as a bomber as well as a patrol/ reconnaissance aircraft and, on several occasions, even served as a 'spotter' aircraft for

Top: PB4Y-2 using JATO units.

Above: Equipped with loudspeakers, a small force of PB4Y-2's was employed in the Marianas towards the end of the war to encourage the surrender of Japanese holed up in caves. After V-J Day the same ships were used over US cities to promote the sale of war bonds. / USN

naval gunfire. Often flying alongside USAAF B-24's, Navy Libs ranged throughout the un-peaceful Pacific on missions that sometimes kept them airborne for fifteen hours or more. Squadrons included VB-101 (ex VP-51), VB-102, VB-104 (second tour as VPB-104), VB-106, VB-108, VB-115, VB-116, VB-117 and VB-200. VP-109 also did a tour in twin-tailed Liberators before being re-equipped with 'Two-By-Fours' as described below.

The Privateer reached operational service in January 1945 when VPB-118 and 119 began operations from Tinian and Midway, respectively. In March VPB-119 moved to Clark Field in the Philippines, from where the Squadron mounted attacks against occupied China until the end of the war, while in April VPB-118 was sent to Okinawa to join the assault against Japan proper. The same month a PB4Y-2B of VPB-109, based on Palawan, launched the first 'Bat' (SWOD-9) guided missile against an enemy target when two were fired at shipping in Balikpapan Harbor, Borneo. This squadron was one of three PB4Y-2B (B for Bat) navy units (the others were VPB-123 and 124) which were equipped with this device before the end of hostilities. Other squadrons operated with improved versions of the Bat after the war. Although each missile was 12 ft long, spanned 10 ft and weighed over 1,600 lb, no internal structural modifications of the original Davis wing were required to accommodate two Bats on operational flights.

Camouflage and Markings

Probably no other aspect of World War II has been the subject of more dedicated research than the camouflage, markings and insignia that appeared on operational aircraft in the combat theatres of that war. Indeed, one can easily be overwhelmed by the profusion of books, pamphlets and magazine articles which cover, in varying degrees of detail, this fascinating subject. This is of course all to the good, for (besides helping to untangle a very tangled subject) it is at the very least a comfort for us to know that, even as millions of men were caught up in a maelstrom of confusion that appeared meaningless at times and often must have seemed impersonal in the extreme, there was a means by which an individual could symbolise and identify a highly personal relationship between himself and his aircraft, his crew, his squadron, his group . . . even his air force.

Once again, therefore, we are going to tread only lightly on an area of well-ploughed ground in favour of the less personal, perhaps less interesting, but certainly more neglected subject of just what Liberators looked like when they first rolled out of those cavernous wombs at San Diego and points east.

LB-30A's and Liberator I's left the Consolidated factory painted to the British contract specification, which was the standard RAF four-engined monoplane pattern executed in Dark Green and a deep brown colour the English termed 'Dark Earth'. All under surfaces were matt black. Wing roundels were type B and applied to upper surfaces only; fuselage roundels were type A.1. Fin flashes, which appeared on both sides of each fin, were the then-current equal tri-colour. Before LIB II deliveries began, the colour scheme was revised to extend the matt black colouring of the under surfaces to include the sides of the fuselage and both sides of each fin and rudder. A portion of the official CAC drawing for LIB II Camouflage Markings & Insignia ('Scheme A') is reproduced on page 172. 'Scheme B' was a mirror image of 'Scheme A'. With the inevitable few exceptions, these variations were used on alternate LIB II aircraft. Most of these aircraft were repainted during subsequent assignments such as the transport role (where black predominated), Coastal Command (Dark Slate Grey with white under surfaces), or in desert service (where 'Middle Stone', a lighter brown, replaced the green).

Early B-24D's destined for the UK (41-11000 range) were specified to retain the then-current USAAF Dark Olive Drab upper surfaces, while the lower surfaces, Neutral Grey on USAAF machines, were to be painted Deep Sky Blue. Wing and fuselage roundel types remained as on the LIB II's, as did the fin flashes. Serial numbers of these aircraft were

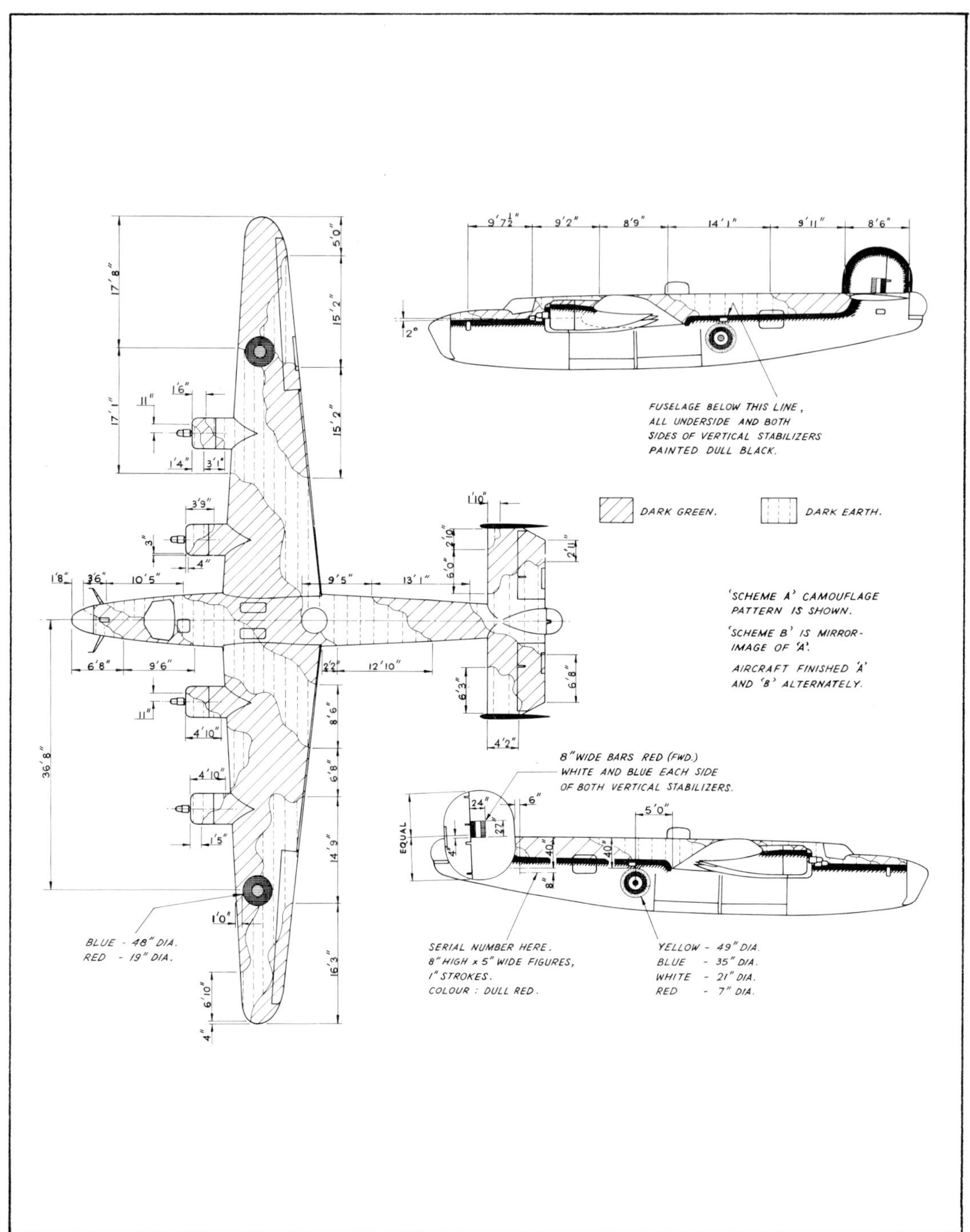

Camouflage and insignia of Liberator II for RAF

relocated to appear on the outside of each fin just ahead of the flash. 'U.S. ARMY', of course, was deleted from the lower wing surfaces. After Pearl Harbor the USAAF took over forty-six of the eighty-three B-24's painted in this colour scheme. Insignia were changed but the Deep Sky Blue under surfaces were not repainted. All of the original HALPRO aircraft, for example, came from this source. Since they differed from all other USAAF machines, they quickly acquired the sobriquet 'blue-bellies'.

Consolidated used Duco (Du Pont Company) paints on all UK Liberators, identified as follows:

Dark Earth—Duco No. 71-035
Dark Green—Duco No. 71-013
Deep Sky Blue—Duco No. 71-052
Dark Olive Drab—Duco No. 71-028
Insignia Yellow—Duco No. 71—010
Insignia Blue—Duco No. 71-038
Insignia White—Duco No. 71-001
Insignia Red—Duco No. 71-011

Lacquer was used in all cases except for the insignia colours, which were in enamel.

After the early B-24D's mentioned above, no special painting was done at the production sites for aircraft destined for Lend Lease allocation except at Fort Worth where, as previously noted, RAF serial numbers were applied as that site became the predominant source of UK-bound Liberators.

Starting with the YB-24 in May 1941, all USAAF B-24 aircraft except the nine B-24A's (see below) received camouflage prior to acceptance in accordance with Army Air Corps Specification 24114, which called for Dark Olive Drab (shade No. 41) on upper surfaces and Neutral Grey (shade No. 43) on under surfaces with the dividing line located 30° below the point where, in level attitude, the aircraft exterior surfaces were vertical. The cross-sections of the Liberator fuselage were such that, in order to maintain a relatively straight line of separation from nose to tail, the 30° specification was handled rather casually. Actual measurements were 35° at the forward edge of the fin, 40° at the leading edge of the waist window, 58° at the forward bomb bay and 48° at the nose section. The Willow Run 'wavy line' method of separating the colours was even more obviously in violation of the specification. Apparently the USAAF never made an issue of it. (Ford's ideas for painting innovations were not limited to the B-24 exterior. At one time the company seriously proposed that the drab interiors of the aircraft be painted in pastel shades which would be varied for each production block—e.g., blue for H-20's, green for H-25's, etc. The USAAF was unimpressed and Ford's B-24's remained unpastelled.)

Bomb bay interiors were Neutral Grey. All propeller blades were black with 4 in. yellow (shade No. 48) tips. The national insignia appeared on left upper and right lower wing surfaces. Of 48 in. diameter until mid-June 1942, it was reduced to 45 in. thereafter. In each case the centre of the star was placed 13 ft 9 in. inboard from the wingtip. Fuselage insignia on camouflaged aircraft was centred at a point 12 ft 6 in. forward of the leading edge of the vertical fin and 55 in. (vertical distance) below the top of the fuselage profile. Originally 64 in. in diameter, the fuselage insignia was reduced to 48 in. during May 1942 and to 45 in. shortly thereafter, becoming the same size as the cocardes on the wings. The red centre disc was removed from the insignia and 'US ARMY' deleted from the wing under surfaces on or about 30 May 1942. This corresponded with the beginning of the assembly of 41-23640. The next insignia change was ordered on 29 June 1943 when the 'star-and-bar' made its appearance. Initially outlined in red, this was changed to a blue outline effective from 17 September 1943.

No evidence has been located that indicates Neutral Grey was ever replaced by Extra Dark Sea Grey (ANA No. 603) as called for by Army Navy Aeronautical Bulletin No. 157, issued in September 1943.

There apparently was some confusion associated with the agreement to give the UK early delivery of some of the USAAF B-24A's in exchange for B-24D's scheduled for later production. While negotiations proceeded as to just how many ships would be involved, the plant production people decided to 'standardise the line' and take care of the discrepancies later. As a result, all twenty-nine Liberators produced in this configuration were painted to British contract specifications. When it was finally decided that twenty of these would go to the RAF as Liberator I's and the balance would be retained by the USAAF as B-24A,s, the USAAF suddenly acquired nine B-24's in British camouflage. The pressures of increasing production precluded time-consuming re-work; so it was that the A's were given a quick insignia change and carried out their globe-spanning adventures while still garbed in RAF Dark Green, Dark Earth and black.

After the deletion of camouflage paint, the fuselage insignia was ordered to be moved forward so that its centre was 37 in. ahead of the waist window and 11 in. below a line projected forward from the lower sill of the waist opening. However, although the placement of wing insignia remained generally consistent with the locations described above, the introduction of multiple Liberator production lines eventually played havoc with the location of the fuselage cocardes. Deviations began with the introduction of the star-and-bar; after this point each manufacturer varied the location just enough, and then revised that location just often enough so as to produce the bewildering total of fifteen variations in all. The manufacturers showed similar individuality in interpreting the specification that the anti-glare panel forward of the pilot's enclosure should 'be of such area and proportion that pilot and co-pilot will not be affected by reflection'.

All exterior stencilling and markings on natural metal finish (NMF) aircraft, including the insignia and anti-glare panel ahead of the windshield, were in enamel paint.

The decision to eliminate camouflage paint on USAAF heavy bombers was based on many factors, the most important of which was the fact that the combination of large formations and enemy radar made it patently impossible to hide strategic air operations from the enemy. Also, by early 1944 the threat of either German or Japanese action against US aircraft parked at their own bases had lessened considerably. In addition, certain advantages were to be gained by deleting paint, e.g., aircraft speed would be increased slightly, weight and production man-hours would decrease and the reflective qualities of unpainted aluminium would make it easier to locate a downed aircraft during search and rescue operations. While in most cases the transition went smoothly, there is some evidence that the sudden appearance over Germany of one complete group (492nd) flying unpainted Liberators was responsible for making it easier for the Luftwaffe to single them out and concentrate on their destruction.

Special purpose painting was usually accomplished at modifications centres—where practically every Liberator ever built (and most every other US warplane as well) was routed directly after delivery. The most famous of these special jobs was the 'desert pink' (or, more properly, shade No. 49 'Sand') series of B-24's turned out in 1942 for service in North Africa and illustrated elsewhere in these pages. It has been noted that the Japanese overture against the Aleutians caused a last-minute diversion of the 404th Bombardment Squadron, which was equipped with these aeroplanes, to the white cold of Alaska. There the 404th Liberators soon became known as *The Pink Elephants.* The name stuck and was still carried on the noses of 404th Liberators, along with a suitable insignia, even after the Squadron received olive drab replacement aircraft. Eventually the joke subsided, however, and by the time the Squadron's official insignia was approved the elephant had changed from pink to white.

During the spring and summer of 1944 many of the Liberators of the Fifth, Eleventh, Thirteenth and Fourteenth Air Forces were field-painted with a high-gloss black paint that had been developed by the Camouflage Section of the National Defense Research Committee. Applied as a top coat, this so-called 'Black Widow' finish produced a remarkably efficient specular reflection when caught in a searchlight beam, rendering the aircraft practically invisible. Moreover, during tests at Eglin Field it was shown to reduce the effectiveness of radar-controlled searchlights by approximately 50%. While unpopular with the combat crews at first, the black finish achieved greater acceptance after the statistics began to show that it *did* do what the scientists claimed it would.

Post-war markings affected very few Liberators, but the few that were still in service dutifully accepted their 'buzz-numbers' which were assigned as follows: BC for B-24's, CR for C-87's and FD for F-7's.

After promising to keep our treatment of unofficial markings to a bare mimimum, the following are offered as representative samples of why the B-24 was sometimes referred to as the 'Flying Billboard'.

NOTES

28 These were the twenty-three B-24D's taken overseas by the Halverson Detachment (HALPRO) as will be described later.

29 The observant reader will have noted by this time that AL606 is unaccounted for after departing Hamilton Field. It is the writer's belief that this aircraft was also lost in the Java campaign, but positive proof is lacking.

30 At later stages of the conflict many other LB-30's found their way into ATC service, including AL570 and AL573—both of which were converted to C-87's.

31 Technically AL610 remained the property of the British; however, it continued to be flown in US markings by Consolidated personnel throughout the war and is therefore not treated as a UK aircraft in any of the totals given in this book.

Above: A very large number of nose paintings were based on a series of pin-up matchbook covers that were in wide circulation during the war. / Via Tom Stutt

Above: Some nose art was modified as time went on . . . or certain local edicts went into effect. / Walter M. Jefferies Collection

Below: Comic strip and motion picture cartoon characters were a favourite subject. / Walter M. Jefferies Collection

Below: Occasionally there was some editorial comment. (*TARFU* **was an advanced condition of** *SNAFU*.) / Walter M. Jefferies Collection

10 February 1945

'Although (the B-24N) may not incorporate all the features . . . found desirable in the Air Force we feel that it is the ultimate in the B-24 series and that any future development for improving it can better be spent, in fact must be spent, on building new aircraft.

'We have just about reached the end of the rope as far as the B-24 is concerned.'

BARNEY M. GILES
Lieutenant General, USA
Chief of Air Staff, Army Air Forces

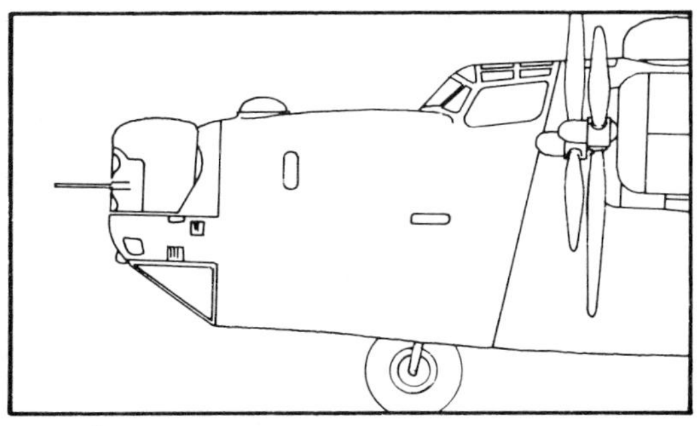

IV Victory and After

The End of Production

Before the end of 1944 Liberator production by Ford at Willow Run had reached such proportions that this facility was able to assume the production quotas of both Douglas and North American. Thus during July and November, respectively, Liberator assembly at Tulsa and Dallas was terminated. Fort Worth rolled out its last B-24 in December 1944 and by VE Day Willow Run and San Diego had begun to phase down. Although the last B-24's from Ford and Consolidated are listed as accepted during June 1945, this was actually an arbitrary contract cut-off point and many of the Liberators in question were actually delivered after that date. By this time, of course, the need for them had long been fulfilled and hundreds of the last Liberators were flown directly to desert storage in the American south-west where they were eventually processed for scrap.

The best index of the USAAF Liberator phase-out is contained in the following figures: during the peak war year of 1944 it operated Liberators a total of 4,115,000 hours. By 1946 the figure was 13,758. During 1947 USAAF B-24's were aloft a mere 451 hours. Most of the 1947 figure was accounted for by the last seven USAAF Liberators to return from war theatre service, these aircraft being flown home from Alaska in April. On 17 September 1947 the B-24 was reclassified from Heavy Bomber to Light Bomber.

The same year saw the RAF phase the GR VIII, its last tactical Liberator, out of the service. Some transport versions lingered on long enough to contribute to the Berlin Airlift in 1948-9 but these, too, were retired soon after. (Scottish Airlines employed two Liberator 'tankers' on the Airlift starting on 19 February 1949. Altogether, these aircraft logged 1,260 hours on the Berlin run.)

As has been noted earlier, the USN continued to operate a few PB4Y-1's along with its Privateers until well into the 1950s, and as recently as 1966 the Indian Air Force maintained a full operational squadron of ex-RAF B-24's as long-range reconnaissance aircraft. The latter fact, incidentally, places the Liberator well up on the list of those World War II aircraft competing for the distinction of being in continuous service for the longest period of time.

The post-war shortage of civilian airliners led to an extension of this phase of the Liberator's career. In February 1947, for example, BOAC had eleven in service. A year later Scottish Aviation was still operating five on its Prestwick to Reykjavik route while, on the other side of the world, two LB 30's still flew bearing the Qantas livery. In fact as late as May 1969 an ex-PB4Y-1, BuNo 90232 (44-42067), is known to have been in active service as a freight hauler under the registration PT-AZX.

But few of the 19,256 built achieved such longevity—except perhaps in the form of aluminium pots and pans. A warplane with no war to fight is not only unneeded and unwanted, but can be a liability as well. This was demonstrated to the USAAF as early as mid-1944 by a B-24D named *Herkimer.* Delivered on 28 November 1942, *Herkimer* first hunted U-Boats with the 18th Antisub

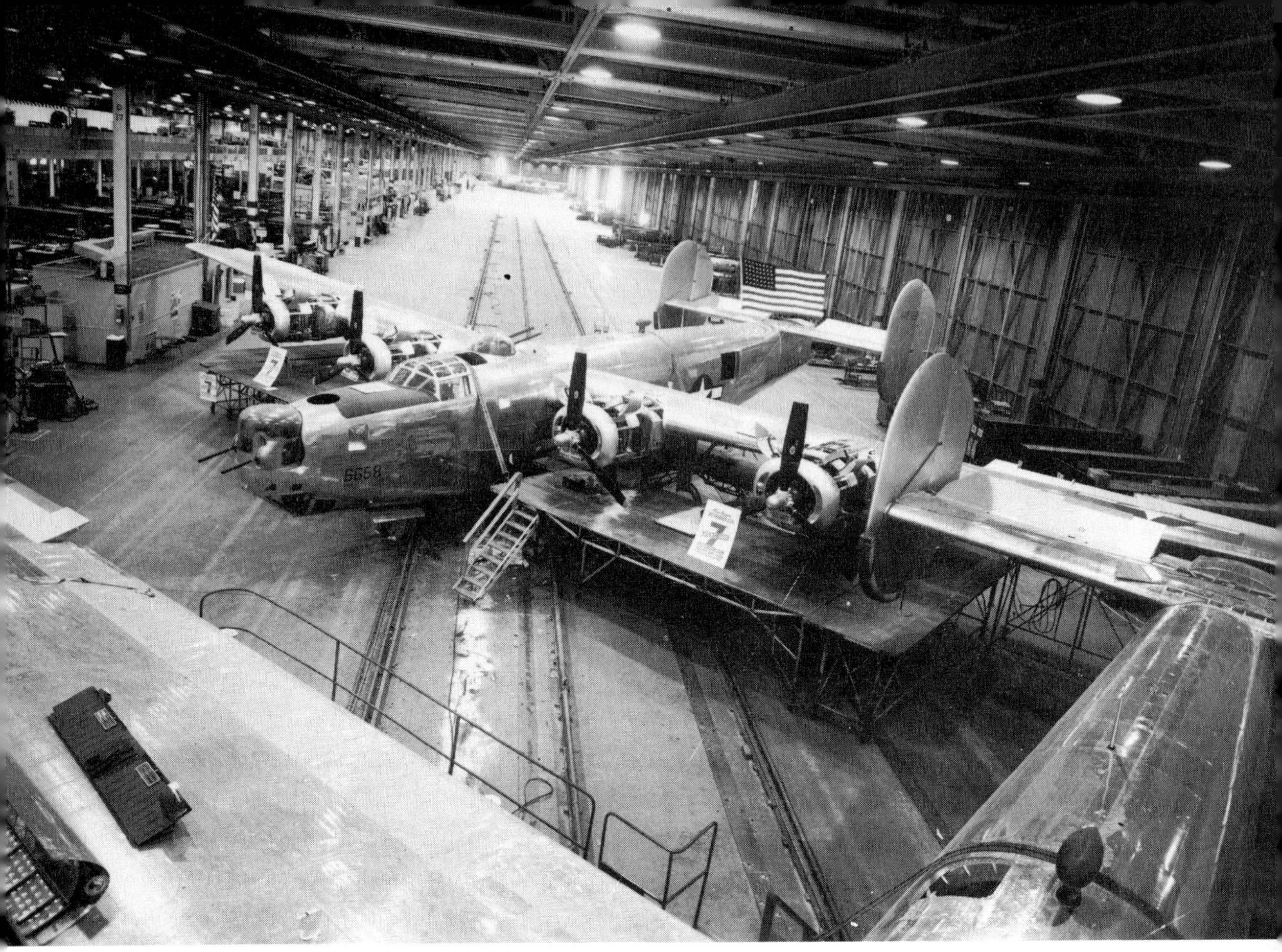
6658

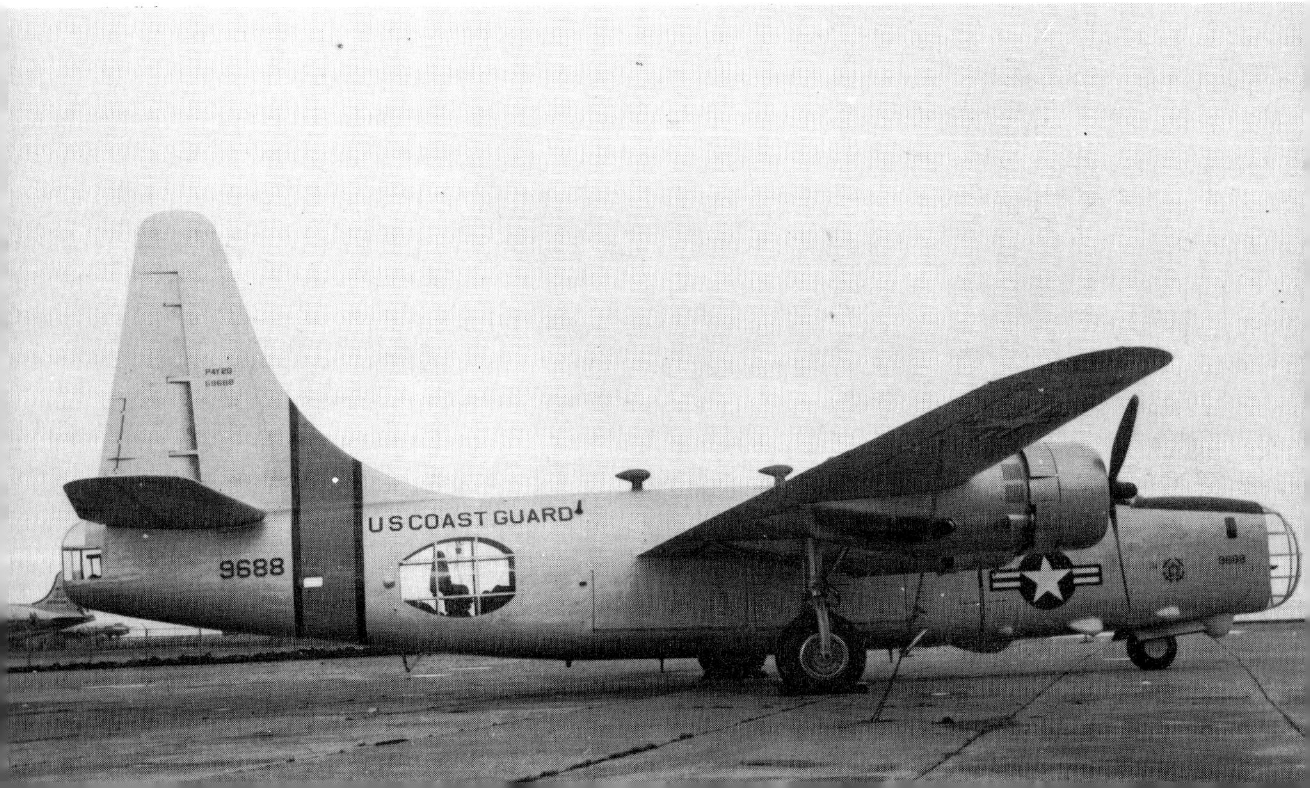
U S COAST GUARD
9688
9688

Left: The last San Diego Liberator moves down the line. It became an F-7B. / General Dynamics

Below left: BuNo 59688 as P4Y-2G. / Via Bob Esposito

Below: Liberator in post-war service with the Indian Air Force. / USAF

Bottom: Liberator II AL552 in post-war commercial dress, 1946. Now G-AHZR, ship was later SX-DAB. / Via Roger Besecker

Squadron. On 19 May 1943 it was transferred to North Africa where it was employed in the bombing role for six months, after which it was flown back to the US and, after a complete overhaul, re-assigned to patrol duties. By 18 July 1944 the ship was an out-and-out war-weary. It was flown to Patterson Field for its last assignment—Air Service Command Project No. 41021. When *Herkimer* was built it had cost the US nearly a third of a million dollars. Project No. 41021 was going to find out just what *Herkimer* was worth now.

After the expenditure of 782·51 man-hours, *Herkimer* was spread over an acre of hangar space, completely disassembled—32,759 lb of aluminium airframe, armament, hydraulics, engines, communications equipment and hundreds of other items. Salvage experts were called in. After poking and prodding at the remains they rendered a unanimous verdict: they didn't want any part of *Herkimer.* Neither did the manufacturers of every conceivable kind of commercial item. Better and cheaper to build our products from scratch, they all said.

So the USAAF removed *Herkimer*'s big, messy carcase from the hangar floor at Patterson, now knowing for sure what they suspected all the time: *Herkimer,* and the thousands of other B-24's just like her, were not worth the expense of salvage.

A few were sold to private corporations (the War Assets Administration's price was $13,750) but the vast majority were melted down. And those that were sold triggered a lawsuit by David R. Davis, who felt his $5.00 per plane royalty agreement with the Government should not apply to Davis-wing aircraft sold as surplus. He tried, unsuccessfully, to collect an additional $795.00 for each plane disposed of in this way.[32]

A few also lingered on at Convair for use as test aircraft, such as B-24J 42-73215 which was modified to accept a General Electric J-35 (TG-180) jet engine as a part of the XB-46 test programme. The jet, housed in the rear fuselage, was fed by an oval-shaped air inlet mounted over the wing similar to the XB-24Q set-up.

Then there was the Model 39, alias the 'XR2Y-1'. To return to this fascinating affair, it will be recalled that this aircraft was owned by CVAC but had been assigned USN BuNo 09803 to allow the company to fly the aircraft pending receipt of a commercial registration from the CAA. This was eventually obtained; the Model 39 then became NX 30039 and BuNo 09803 was retired to oblivion. The second aircraft (the one purchased unfinished from the Navy) was completed and registered as NX 3939. By this time, with 39's all over the aircraft, CVAC had changed the model designation to CVAC Model 104 and was attempting to sell the design as a forty-eight passenger transport. When no orders were forthcoming, NX 30039 was rented to American Airlines which tried it out as a cargo carrier but this experiment, too, produced no production order. The second aircraft, NX 3939, was still assigned to the CVAC Flight Research Department (where it was undergoing tests with full span flaps) when, on 17 September 1945, the company ordered both aircraft scrapped.

And so the Consolidated Liberator passed, first through limbo and then on into history and not a dry eye in the house. But it was there when it was needed. In numbers exceeding any other US warplane. In every theatre of operations. In every square mile of enemy sky—until there *was* no sky that didn't belong to it and its contemporaries. Certainly no other aircraft deserves a greater share of credit for that victory.

Liberator vs Fortress—A Comparison

It is not generally realised just how favoured a position the B-24 occupied relative to the B-17 during the period when the US did its first large-scale shopping for the tools of war. But the cold hard evidence—represented by the Army's willingness to lay out cold hard cash to buy aeroplanes—makes it clear that the Consolidated Liberator was to be the mainstay of the Air Corps' heavy bombardment forces. By the spring of 1941, with a grand total of two B-24's in its inventory, the Army had already placed orders for over 2,350 Liberators while Fortress contracts totalled less than 950. The large Government-financed aircraft plants

Top: *Strawberry Bitch* (42-72843) takes off from Davis-Monthan AFB, Arizona after having been rescued from oblivion and completely restored. Photo was taken on 12 May 1959 as the aircraft left Tucson for the Air Force Museum at Wright-Patterson AFB, Ohio, where it is now on display. This was almost certainly the last flight of a B-24D. Pilot was Col Albert Shower.

Above: Nose of the *Bitch* was at one time painted to represent the 98th Bomb Group Liberator flown by Col John R. Kane on the 1 August 1943 Ploesti mission. / Ken Sumney

Top: This C-87A was acquired after the war by millionaire Milton Reynolds. Ship is shown here at Haneda Air Base, Japan, prior to being used in an attempt to measure the height of Mt. Amne Machine in Western China. / US Army

Above: Dallas, Texas, 17 March 1956. / Via Roger Besecker

Above right: The Model 39 / 104 (Passenger version). / Consolidated

Right: B-24J-CF 44-44175, ex-RAF KH304, ex-Indian Air Force HE 877, returns to Tucson for display as N7866. Date: 27 April 1969. / USAF

09803
LIBERATOR-LINER

planned for Fort Worth, Dallas and Tulsa, which originally had been scheduled to produce B-17's, were shifted to B-24 production while the prestigious Ford Motor Company, first of the giant car makers to be brought into the aircraft programme, was likewise assigned to the Liberator effort.

The decision to go with the B-24 was well justified at the time. The early performance estimates of the Liberator were being realised and in nearly every category they exceeded those of the B-17. The British liked the aircraft and were pressing for what seemed like impossible numbers of them. The file of favourable reports by Army pilots who had flown the Liberator continued to grow. Without question, Consolidated had come up with a winner.

There were certain other contrasts between the two aircraft which were becoming apparent to men who were acquainted with both. For example the overall impression one got from the Fortress interior was that it was, like its exterior, round and smooth—with its equipment built-in rather than added-on. Each B-17 crew member had a place to sit down and strap himself in—a small point, perhaps, but psychologically important. On the other hand the Liberator fuselage, while of larger dimensions than the Fortress, offered little in the way of comfort for the crew. There seemed to be draughts everywhere, and of such magnitude that they were far more than the troublesome spot heaters could contend with. Movement throughout the ship was awkward and difficult in full flight gear, and more often than not resulted in jarring collisions with various sharp-edged and unyielding structural members and/or installed equipment. Idle gunners sat on the floor—if they sat—and likely as not pondered possible fates for the design engineer who was responsible for a fuel-transfer system that required any prudent B-24 pilot to crack open the bomb bay doors in flight to disperse the petrol fumes. Or perhaps the ball and tail gunners thought about the greater speed with which their B-17 counterparts could exit their stations in an emergency.

In any case, the contrast in creature comfort was considerable. It stemmed, in large part, from the different philosphies represented by an aircraft designed and refined in peacetime and one swiftly conceived under the skies of war. But it was performance characteristics that interested the high-level planners, to whom such statistics as range and bomb load were all-important. And, inasmuch as these men were apt to have more influence on procurement matters than tail gunners, it was the Liberator that got the nod.[33]

The first truly mass-produced Liberator, of course, was the B-24D. This was the contemporary of the B-17F Flying Fortress and, at that stage in the development of the two aircraft, the Liberator continued to possess a definite edge over its rival in terms of operational performance. At 32,600 lb empty, the basic weight of the early B-24D was some 1,400 lb lighter than the B-17F while its greater maximum take-off weight—60,000 lb versus 56,500 lb—resulted in a useful load capacity that exceeded the B-17's by nearly 2½ tons. The Liberator was also slightly faster and, because of this plus its greater load capability, could transport a similar bomb load to a far greater range than the Fortress, simply by carrying more fuel for its utterly dependable Pratt & Whitney engines. At equal gross weights the wing loading of the Liberator was over 35% greater than the Fortress, and ceiling was not as high as the B-17. But it was not low enough to cause concern—at least, not yet. Far more important at that moment was a development phase long since passed by the Fortress. The Liberator was being assigned in large numbers to the Training Command and it was getting a reputation as a man-killer.

During World War II it was not unusual for a new warplane, particularly one that advanced the state-of-the-art, to acquire a bad reputation during the course of its introduction to service use. The P-38 Lightning, P-47 Thunderbolt, B-26 Marauder and B-29 Superfortress are cases in point. All of these high-performance aircraft emerged from somewhat shaky beginnings to become outstanding and dependable performers, inspiring the confidence of nearly all who flew them.

21
21

The Liberator likewise suffered as thousands of new crews began to master the weapon that they were to take into combat. The results generated the following comment in the official Air Force history of World War II:

> 'During 1942 and 1943 the B-24 had the sorry distinction of being the Second Air Force's "problem plane". In 1943 alone, 850 Second Air Force combat crewmen went to their deaths in 298 B-24 accidents.'

Why the Liberator should have been singled out for such special mention in the Air Force history is hard to understand. The only really valid statistic here is accident *rate*, i.e.—how many accidents a particular model suffered in relation to its frequency of use, usually expressed in accidents per 100,000 flying hours. By this criterion, in 1943, the year cited above, *only C-47's and Basic Trainers had a lower US accident rate than the B-24*. The figures follow:

Accidents per 100,000 Flying Hours in Continental US

All BT's	31	C-46	59	P-47	163
C-47	31	All AT's	64	P-38	165
B-24	39	B-26	65	P-51	210
B-17	39	B-29	72	P-39	228
B-25	44	All L-Type	130	A-36	273
All PT's	54	A-20	155	P-40	297

Regardless of the perspective the problem gains in retrospect, it was real at the time. As with the other aircraft that suffered similar setbacks, the problem was cured by substituting confidence for carelessness and mastery for mistrust. By the time the Liberator groups were being deployed in large numbers there was no general lack of confidence in the aircraft *per se*. In fact, as far as confidence in their aircraft was concerned, the attitude of most combat crews was keyed to their pilot. If he was the master of his machine, confident in its ability to fly and his ability to fly it, this confidence was readily transferred to the crew and there was seldom a problem.

Inevitably the Liberator continued to be compared with the Flying Fortress and, as far as the later versions of the two bombers were concerned, suffered by the comparison. This was due not so much to short-comings of the Liberator—although it certainly had them—as to the fact that the B-17 was in many respects an exceptional aircraft, with many of its merits having particular—and personal—appeal to the men who flew it and who flew in it. From the beginning the Fortress was an honest aircraft, easy to fly in formation, with a low landing speed and no major vices. 'A four-engined Piper Cub' was the popular and rather apt description. Most important, the Fortress retained its original characteristics throughout its development while the Liberator did not. The B-17, although it was by no means as invincible as its wartime press clippings would indicate, was of rugged construction and withstood battle damage well. While slower than the Liberator, it could fly higher and this factor alone could well have accounted for its favoured status in Europe. After all, there was no substitute for altitude when it came to flak, particularly if there was a formation of Liberators 2,000-3,000 ft below you offering better targets for the ground gunners to shoot at. The higher-flying Fortress crews were quick to admit that a formation of Liberators was 'the best escort we could have'.

The same remark held true for fighter opposition. For while there are no valid statistics that will substantiate the widely-quoted opinions that the Liberator was more susceptible to minor damage and more easily set on fire, the important fact is that a majority of *Luftwaffe fighter pilots* apparently believed this to be true and, consequently, would attack a B-24 rather than a B-17 if given a clear choice.

Actually the Liberator *never did* lose its performance edge over the B-17, as a series of tests run at Eglin Field demonstrated conclusively late in the war. Rather, the areas in which the B-24 excelled became less important in the European and Mediterranean theatres. The range of the Fortress was adequate for Europe, and individual aircraft speed became academic because of formation requirements. Altitude, however, became paramount and here, literally, the B-17 remained on top. In addition, with over 70% of Eighth Air Force mission failures being attributed to navigational errors, the superior accommodations of the B-17 nose were highly desirable. General Doolittle, in fact, considered poor visibility the number one fault of the B-24.

In the Pacific there was no vocal contest between the two aircraft, for although B-24's were originally requested by Pacific theatre commanders because they felt there was a better chance of getting them than the more popular B-17's, the Liberator's longer legs soon demonstrated that it was a natural choice for an air war conducted for the most part at extreme range. The European requirement for tight formation flying was not as severe, and the typical maximum-range mission allowed Pacific Liberators, when necessary, to approach the target at adequate altitude because of the large amount of fuel burned on the way.

Six months after Pearl Harbor it was stated USAAF policy 'to replace B-17 type aircraft with B-24 type aircraft in all of the combat theatres throughout the world except the United Kingdom'. While this plan was never completely realised (six Fortress groups remained with the Fifteenth Air Force and twelve Liberator groups stayed with the Eighth) the total number of USAAF first line B-24's exceeded the equivalent B-17 figure by the end of January 1944 (3,980 vs 3,717), and six months later Liberator figures topped the Fortress by nearly 1,400 (5,906 vs 4,525). This numerical superiority continued to edge higher until it stood at 1,484 at the end of April 1945. Actual commitments to overseas theatres saw an even greater spread in favour of the B-24—the difference reaching over 1,600 aircraft by 1 August 1944.

With regard to actual bombing efficiency, the US Strategic Bombing Survey made a brief comparison of B-17's and B-24's with regard to bombing results. This study was summarised in part as follows:

> 'A comparison of the accuracy achieved by the B-17's and B-24's within the Eighth Air Force . . . (shows) . . . that the B-17's averaged 40·77 per cent of their bombs within a thousand feet of the aiming point, while B-24's averaged 37·8 per cent. In . . . making the better record, the B-17's suffered, on the average, by attacking at a higher altitude with a larger attacking force and with more aircraft per box than the B-24's. Hence, there is no doubt but that the B-17 was a heavy bomber superior to the B-24.'

This summary fails to point out, however, that the statistics used to arrive at this conclusion were chosen to include only the period when both types were active simultaneously, and the B-17 went operational in England before the B-24. Thus, for the Fortress the data covering the important breaking-in period (a time usually characterised by generally poor bombing) is excluded, while the comparison *does* include initial Eighth Air Force Liberator operations. If initial B-17 operations *are* counted, Fortress accuracy drops off to 33·6%, or less than the quoted figure for the B-24.

Of even greater significance, however, is that the conclusion overlooks the fact that, beginning in October 1944 and for every month thereafter until the end of the war (except for December 1944 when there was a virtual tie) B-24 accuracy was better than the B-17 by as much as 20%. To ignore this obvious trend while rating the B-17 superior to the B-24 as a bomber seems somewhat short-sighted.

In any case, the most important point to remember is what was accomplished with the Liberator *in spite of* its limitations. And this, of course, becomes the ultimate tribute not to the aircraft itself but to the men who flew and fought in the machine they called 'the box the B-17 came in'. Writing in their own historical narrative shortly before the end of the Pacific War, one group expressed this interdependence of man and machine in the following way:

> 'The Liberator has, without seeming to, ruled our lives for three years. Indirectly, we have been the slaves of the Flying Boxcar. She has been the taskmaster, determining when we go to work, and when we quit. She can get you up in the middle of the night, or early in the morning, keep you working away, or she can let you hit your sack. Loving

her, cursing her, berating her, and again praying for her, you become her duty-bound slave.

'Human-like, the B-24 is rugged and yet sensitive, crude looking and yet beautiful, clumsy, and yet clean-lined. She is at all times temperamental. No two B-24's are alike. For weeks she can be gay and healthy, the engines sing and zoom, then, without warning, she can become unruly, her instruments register wrong, her controls become sluggish, her engines spit oil. One plane can hungrily lap up gas, another sips it lady-like, another can tear away and fly faster than her sisters, while still another can only snail along. One plane is wing weighted, another is tail heavy. A B-24 has her moods. She needs attention; at times the attention of a squawking baby, at others she gaily bounces along, cat-purring with contentment.

'Our history. . .is a saga of the B-24. The men of this Group have nursed every ounce of energy out of her. We have taken her on the longest formation flights of the war. We have taken her through thick ack-ack, and with her we have thumbed our noses at the Nip Air Force and have rid the air of a goodly number of Jap planes. With her, we have taken on the Jap Navy, and slugged the hell out of it. With her, we have pock-marked acre after acre of the Japanese real estate, and destroyed village after village, town after town, and have driven the Jap away from his cities. The B-24 has been our weapon—our rifle—our fighting tool.'

It was indeed.

NOTES

32 In doing so, Davis lost ground; in the legal battle that followed, the US Court of Claims ruled that twelve out of the thirteen claims of the original Davis Patent were invalid.

33 At least one B-17 crew is known to have stood in awe of the Lib's performance. Eastbound one morning about an hour out of Gander, Newfoundland, a B-24 replacement crew enroute overseas spotted a Fortress ahead and below. Pouring on the coal and going into a shallow dive, the Lib pilot built up airspeed until he was closing on the Fortress at a maximum rate. At this point he levelled off, shut down the two outboard engines, feathered the props and went sailing by the startled B-17 crew. That night at Gander the latter were still shaking their heads over 'that damned B-24 that passed us on only two engines'.

LB-30B AM927 was operated by the Continental Can Company as N1503, later as XC-CAY, and finally in the livery of the Confederate Air Force in 1970. Without question, AM927 is today's oldest surviving Liberator. / Confederate Air Force

Appendices

Appendix A: Summary of Liberator/Privateer production contracts

	Contract		Number of Aircraft:			
Date	*Number*	*Site*	*Ordered*	*Built*	*Trans. (T) or Canc. (C)*	**Remarks**
3/39	12436	CO	1	1	—	1 XB-24
4/39	12464	CO	7	7	—	1 YB-24, 6 B-24D
	A-5068	CO	6	6	—	6 LB-30A
	F-677	CO	159	160	—	20 LIB I, 140 LIB II
9/39	13281	CO	38	38	—	9 B-24A, 9 B-24C, 20 B-24D
8/40	13281-5	CO	56	56	—	56 B-24D
9/40	16005	CO	352	305	47(T)	305 B-24D, See footnotes for transfers
5/41	DA-4	CO	700	629	71(T)	629 B-24D, See footnotes for transfers
5/41	18722	DT	600	600	—	167 B-24E, 433 B-24H
9/41	21216	FO	795	795	—	490 B-24E, 305 B-24H
9/41	18723-1	CF	800	232	568(T)	175 B-24H, 57 B-24J, 568 transferred to FO -32
2/42	24620	CO	1,200	1,200	—	1,200 B-24D
4/42	21216-4	FO	700	700	—	700 B-24H
4/42	18722-1	DT	1,000	354	646(T)	149 B-24H, 205 B-24J, 646 transferred to FO -32
5/42	24663	NT	750	430	320(C)	430 B-24G
5/42	26992	CF	750	750	—	295 B-24D, 37 B-24E, 70 B-24H, 348 B-24J
6/42	18723-3	CF	8	8	—	8 B-24D[1]
6/42	18722-4	DT	10	10	—	10 B-24D[2]
7/42	30461	CO	750	750	—	199 B-24D, 551 B-24J
8/42	21216-5	FO	900	900	—	775 B-24H, 125 B-24J
11/42	35312	CO	900	900	—	900 B-24J
11/42	26992-1	CF	200	200	—	200 B-24J
4/43	18723-7	CF	180	180	—	169 C-87[3], 6 C-87A[4], 5 AT-22[5]
12/43	24663-10	NT	650	536	114(C)	536 B-24J
12/43	40033	CO	4,500	2,674	1,373(C)	1,341 B-24J, 417 B-24L, 916 B-24M
	40033	CF		453		453 B-24J
1/44	21216-24	FO	4,100	3,183	917(C)	248 B-24J, 1,250 B-24L, 1,677 B-24M, 8 B-24N
2/44	40715	CF	500	500	—	500 B-24J
	1675(Ltr)	CO	125	—	125(C)	Planned C-87C production
3/44	811	CF	111	111	—	111 C-87
6/44	21216-32	FO	2,214	1,214	1,000(C)	1,214 B-24J
3/45	21216-47	FO	2,175	—	2,175(C)	Planned B-24N production
10/43	NOa(s)855	CO	660	660	—	660 PB4Y-2
3/44	NOa(s)3236	CO	112	34	78(C)	34 RY-3
10/44	NOa(s)4841	CO	710	80	630(C)	80 PB4Y-2
			27,319	19,256	1,332(T) 6,732(C)	

[1]All from contract 16005
[2]Includes 4 from 16005 and 6 from DA-4
[3]Includes 35 from 16005 and 61 from DA-4
[4]Includes 3 from DA-4
[5]Includes 1 from DA-4

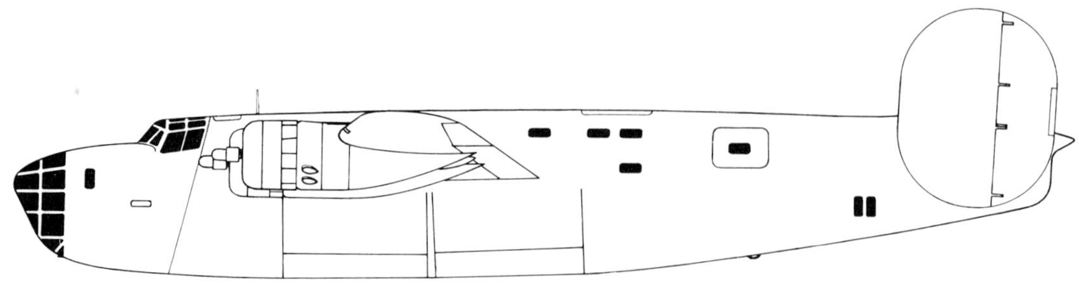

XB-24

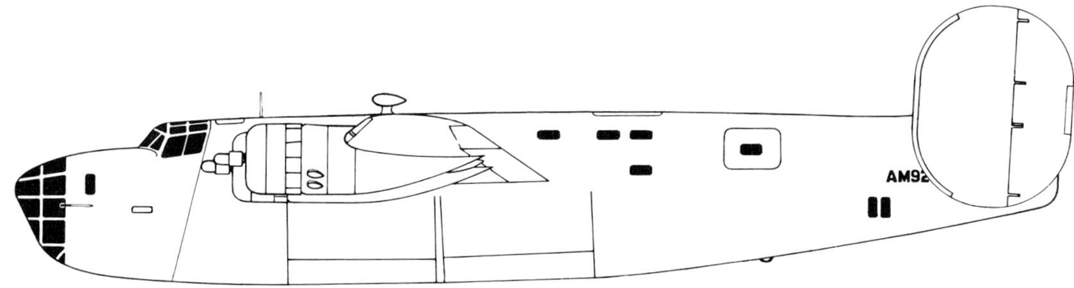

Liberator I

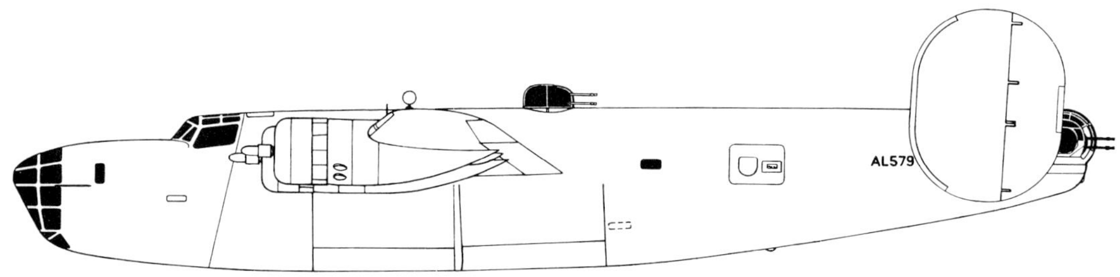

Liberator II

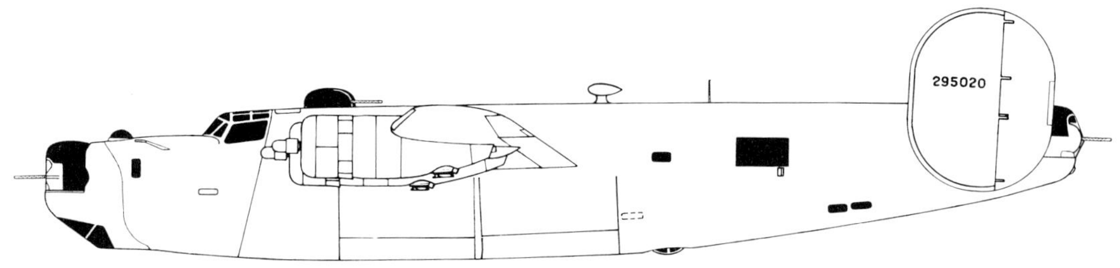

B-24H-FO

Norman Ottaway

Appendix B: Liberator/Privateer production by Plant Location, by Type, Model and Series

	CO	*CF*	*FO*	*DT*	*NT*	*Total*
XB-24	1	—	—	—	—	1
LB-30A	6	—	—	—	—	6
LIB I	20	—	—	—	—	20
YB-24	1	—	—	—	—	1
B-24A	9	—	—	—	—	9
LIB II	140	—	—	—	—	140
B-24C	9	—	—	—	—	9
B-24D	2,415	303	—	10	—	2,728
B-24E	—	144	490	167	—	801
B-24G	—	—	—	—	430	430
B-24H	—	738	1,780	582	—	3,100
B-24J	2,792	1,558	1,587	205	536	6,678
B-24L	417	—	1,250	—	—	1,667
B-24M	916	—	1,677	—	—	2,593
XB-24N	—	—	1	—	—	1
YB-24N	—	—	7	—	—	7
C-87	—	280	—	—	—	280
C-87A	—	6	—	—	—	6
AT-22	—	5	—	—	—	5
PB4Y-2	740	—	—	—	—	740
RY-3	34	—	—	—	—	34
Total	7,500	3,034	6,792	964	966	19,256

Appendix C: Liberator acceptances by Month—by Manufacturer★

		J	*F*	*M*	*A*	*M*	*J*	*J*	*A*	*S*	*O*	*N*	*D*	*Total*
1940	CO	—	—	—	—	—	—		1	—	—	—	6	/
1941	CO	—	—	2	6	12	9	1	19	27	30	30	33	169
1942	CO	12	59	71	80	86	95	100	108	116	129	120	147	1,123
	CF	—	—	—	1	2	2	3	—	9	8	13	12	50
	DT	—	—	—	—	—	—	—	2	—	1	5	—	8
	FO	—	—	—	—	—	—	—	—	2	1	10	11	24
		12	59	71	81	88	97	103	110	127	139	148	170	1,205
1943	CO	120	148	152	162	177	190	202	214	230	245	250	255	2,345
	CF	14	24	35	50	59	85	125	162	134	178	162	205	1,233
	DT	2	5	3	3	15	30	39	45	50	53	78	71	394
	FO	14	36	60	95	115	107	84	116	149	163	159	193	1,291
	NT	—	—	1	—	—	2	2	4	6	12	13	21	61
		150	213	251	310	366	414	452	541	569	651	662	745	5,324

1944	CO	253	254	270	251	270	260	169	105	212	168	130	108	2,450
	CF	230	216	204	172	192	129	124	117	104	104	94	65	1,751
	DT	70	94	78	78	78	87	77	—	—	—	—	—	562
	FO	217	238	309	326	350	383	415	428	377	340	312	296	3,991
	NT	55	66	90	84	95	109	100	100	90	85	31	—	905
		825	868	951	911	985	968	885	750	783	697	567	469	9,659
1945	CO	103	99	126	130	102	71	—	—	—	—	—	—	631
	FO	324	296	310	252	179	125	—	—	—	—	—	—	1,486
		427	395	436	382	281	196	—	—	—	—	—	—	2,117
														18,481

★ Excludes production PB4Y-2's and RY-3's. Also excludes AL503, which crashed before acceptance.

Appendix D: Liberator / Privateer production Time Line

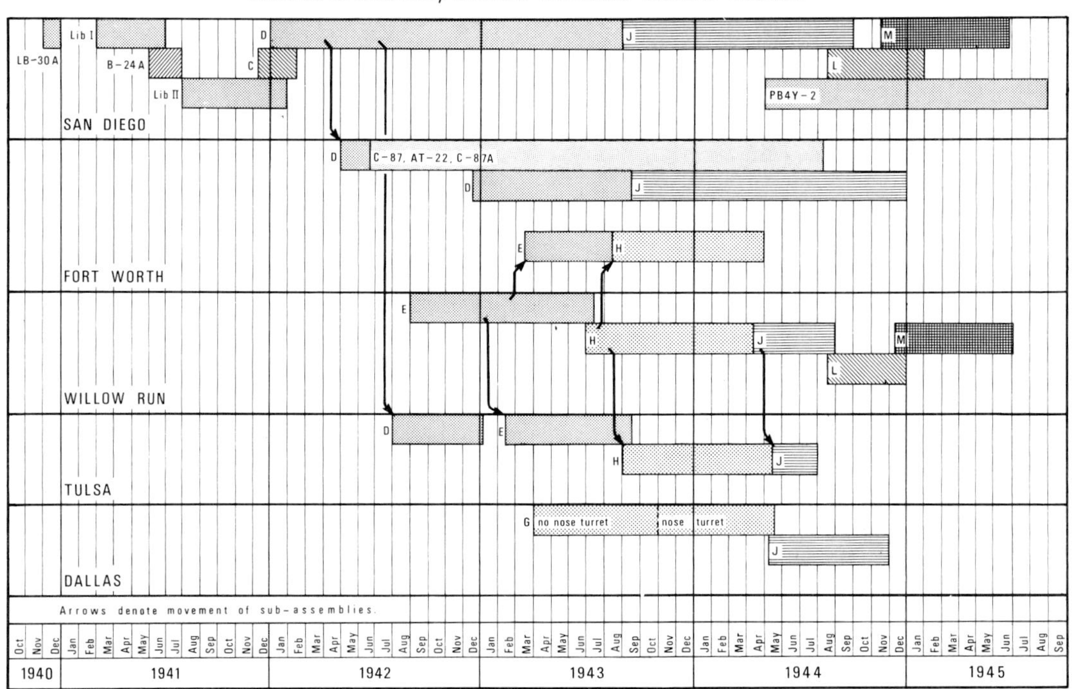

440052

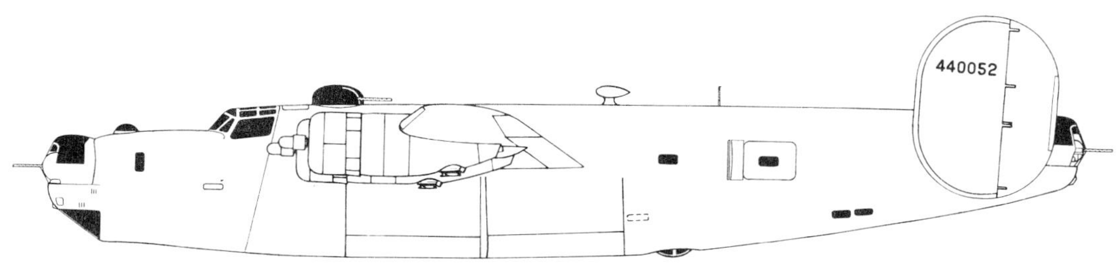

B-24J-CO

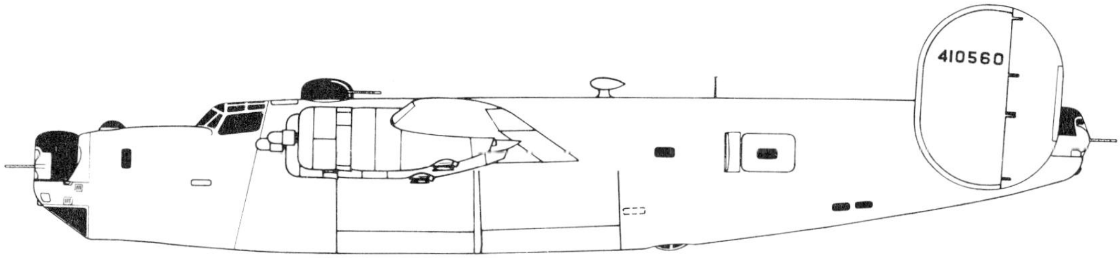

410560

B-24J-CF

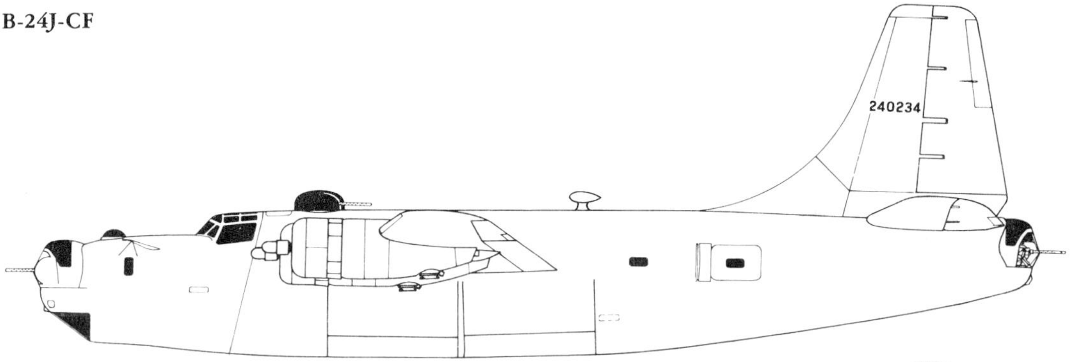

240234

XB-24K

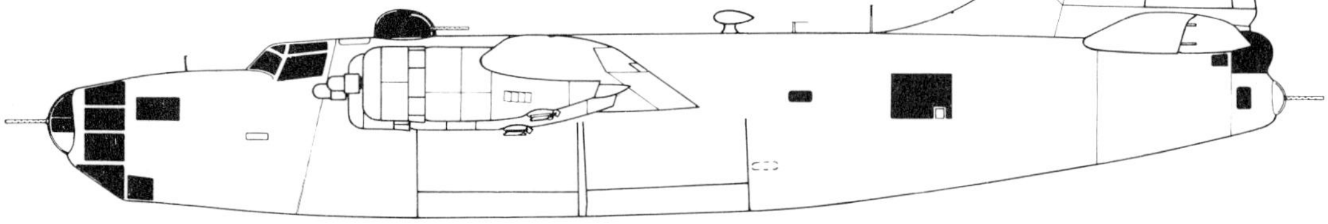

448753

YB-24N

Norman Ottaway

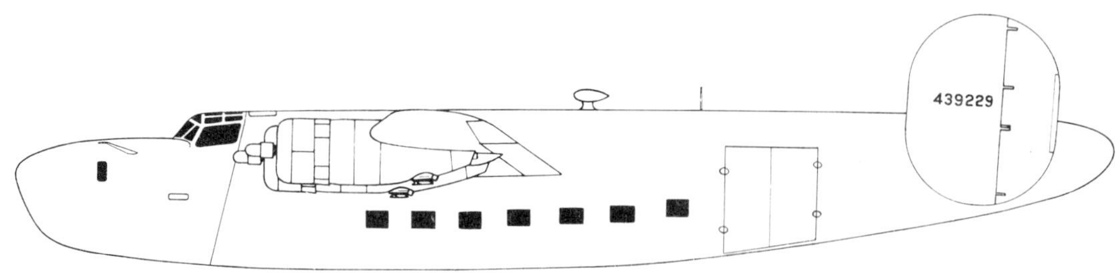

C-87

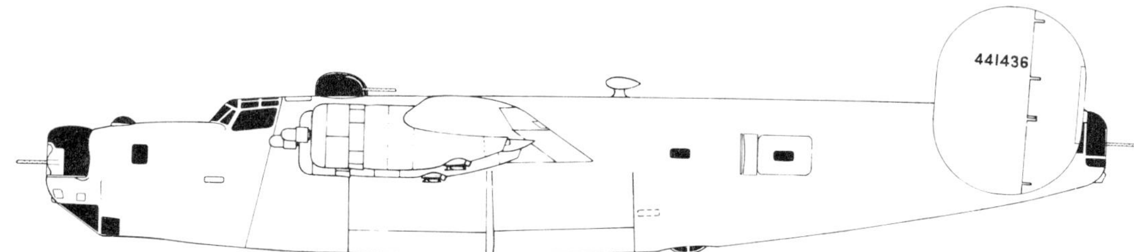

B-24L-CO

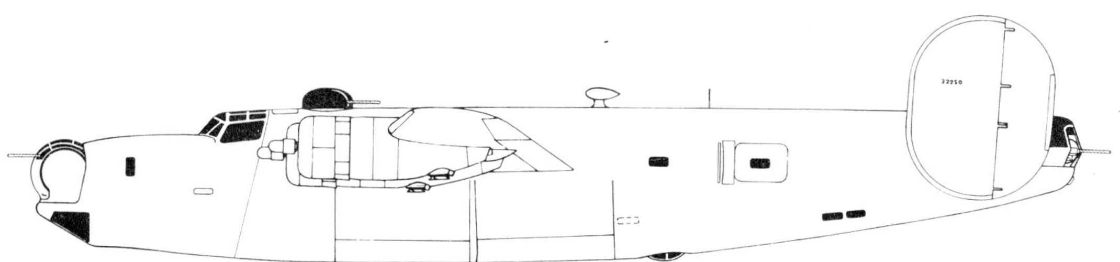

PB4Y-1

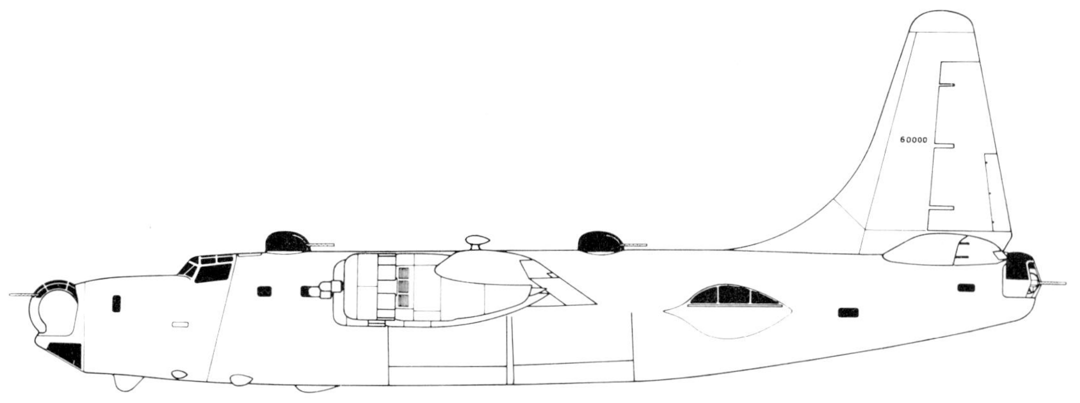

PB4Y-2

Norman Ottaway

Appendix E: Numerical List of Liberator/Privateer serial numbers

Army Cognisance Aircraft

Serial Range		*Number*	*Type/Model*	
AL503	AL641	139	LIB II	
AM258	AM263	6	LB-30A	
AM910	AM929	20	LIB I	
FP685		1	LIB II	
39-680		1	XB-24	(XB-24B)
40-696	40-701	6	B-24D	(CO)
40-702		1	YB-24	(CO)
40-2349	40-2368	20	B-24D	(CO)
40-2369	40-2377	9	B-24A	(CO)
40-2378	40-2386	9	B-24C	(CO)
41-1087	41-1142	56	B-24D	(CO)
41-11587		1	B-24D	(CO)
41-11588	41-11589	2	B-24D	(CF)
41-11590	41-11603	14	B-24D	(CO)
41-11604	41-11605	2	B-24D	(CF)
41-11606		1	B-24D	(CO)
41-11607		1	B-24D	(CF)
41-11608		1	C-87	(CF)
41-11609	41-11626	18	B-24D	(CO)
41-11627	41-11628	2	B-24D	(CF)
41-11629	41-11638	10	B-24D	(CO)
41-11639	41-11642	4	C-87	(CF)
41-11643	41-11654	12	B-24D	(CO)
41-11655	41-11657	3	C-87	(CF)
41-11658	41-11673	16	B-24D	(CO)
41-11674	41-11676	3	C-87	(CF)
41-11677	41-11703	27	B-24D	(CO)
41-11704		1	C-87	(CF)
41-11705		1	B-24D	(CF)
41-11706	41-11709	4	C-87	(CF)
41-11710	41-11727	18	B-24D	(CO)
41-11728	41-11733	6	C-87	(CF)
41-11734	41-11741	8	B-24D	(CO)
41-11742	41-11747	6	C-87	(CF)
41-11748	41-11753	6	B-24D	(CO)
41-11754	41-11756	3	B-24D	(DT)
41-11757	41-11787	31	B-24D	(CO)
41-11788	41-11789	2	C-87	(CF)
41-11790	41-11799	10	B-24D	(CO)
41-11800		1	C-87	(CF)
41-11801	41-11836	36	B-24D	(CO)
41-11837	41-11838	2	C-87	(CF)
41-11839	41-11863	25	B-24D	(CO)
41-11864		1	B-24D	(DT)
41-11865	41-11906	42	B-24D	(CO)
41-11907	41-11908	2	C-87	(CF)
41-11909	41-11938	30	B-24D	(CO)
41-23640	41-23668	29	B-24D	(CO)
41-23669	41-23670	2	C-87	(CF)
41-23671	41-23693	23	B-24D	(CO)
41-23694	41-23696	3	C-87	(CF)
41-23697	41-23724	28	B-24D	(CO)
41-23725	41-23727	3	B-24D	(DT)
41-23728	41-23755	28	B-24D	(CO)
41-23756	41-23758	3	B-24D	(DT)
41-23759	41-23790	32	B-24D	(CO)
41-23791	41-23793	3	C-87	(CF)
41-23794	41-23849	56	B-24D	(CO)
41-23850	41-23852	3	C-87	(CF)
41-23853	41-23858	6	B-24D	(CO)
41-23859	41-23862	4	C-87	(CF)
41-23863		1	C-87A	(CF)
41-23864	41-23902	39	B-24D	(CO)
41-23903	41-23905	3	C-87	(CF)
41-23906	41-23958	53	B-24D	(CO)
41-23959		1	C-87	(CF)
41-23960	41-24003	44	B-24D	(CO)
41-24004	41-24006	3	C-87	(CF)
41-24007	41-24026	20	B-24D	(CO)
41-24027	41-24029	3	C-87	(CF)
41-24030	41-24138	109	B-24D	(CO)
41-24139	41-24141	3	C-87	(CF)
41-24142	41-24157	16	B-24D	(CO)
41-24158		1	C-87	(CF)
41-24159		1	C-87A	(CF)
41-24160	41-24163	4	C-87	(CF)
41-24164	41-24171	8	B-24D	(CO)
41-24172	41-24173	2	C-87	(CF)
41-24174		1	C-87A	(CF)
41-24175	41-24311	137	B-24D	(CO)
41-24339		1	B-24D	(CO)
41-28409	41-28573	165	B-24E	(DT)
41-28574	41-29006	433	B-24H	(DT)
41-29007	41-29008	2	B-24E	(DT)
41-29009	41-29115	107	B-24E	(CF)
41-29116	41-29608	493	B-24H	(CF)
42-6976	42-7464	489	B-24E	(FO)
42-7465	42-7769	305	B-24H	(FO)
42-7770		1	B-24E	(FO)
42-40058	42-41257	1,200	B-24D	(CO)
42-50277	42-50451	175	B-24H	(CF)
42-50452	42-50508	57	B-24J	(CF)
42-50509	42-51076	568	B-24J	(FO)
42-51077	42-51225	149	B-24H	(DT)
42-51226	42-51430	205	B-24J	(DT)
42-51431	42-52076	646	B-24J	(FO)
42-52077	42-52776	700	B-24H	(FO)
42-63752	42-64046	295	B-24D	(CF)
42-64047	42-64394	348	B-24J	(CF)
42-64395	42-64431	37	B-24E	(CF)
42-64432	42-64501	70	B-24H	(CF)
42-72765	42-72963	199	B-24D	(CO)
42-72964	42-73514	551	B-24J	(CO)
42-78045	42-78474	430	B-24G	(NT)
42-78475	42-78794	320	B-24J	(NT)
42-94729	42-95503	775	B-24H	(FO)
42-95504	42-95628	125	B-24J	(FO)
42-99736	42-99935	200	B-24J	(CF)
42-99936	42-100435	500	B-24J	(CO)
42-107249	42-107265	17	C-87	(CF)

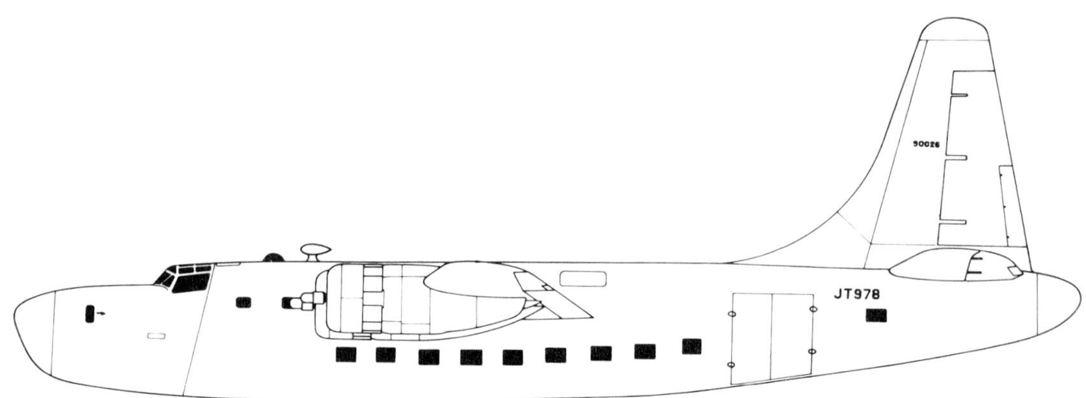

Liberator CIX (RY-3)

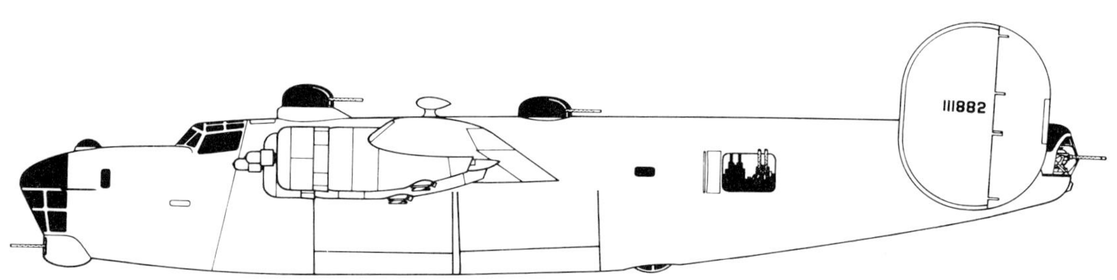

XB-41

STATION DIMENSIONS GIVEN IN INCHES FROM STA. O

0 55 94 130 151 189 224 263 309 353 397 438 484 528 572 616 661 698¼ 736 791½

36 74½ 112 147⅜ 172 206 244 286 331 375 419 461 506 550 594 639 683 714 751⅞

STA. 0·1 0·2 0·3 1·1 1·2 2·1 3·1 3·2 4·1 4·2 4·3 4·4 5·1 5·2 5·3 5·4 6·1 6·2 7·1 7·2 7·3 7·4 7·5 7·6 7·7 8·1 9·1 9·2

STA.O 1·0 2·0 3·0 4·0 5·0 6·0 7·0 8·0 9·0 10·0

B-24D Fuselage stations

Serial Range		Number	Type/Model	
42-107266		1	AT-22	(CF)
42-107267	42-107275	9	C-87	(CF)
42-109789	42-110188	400	B-24J	(CO)
43-30548		1	C-87	(CF)
43-30549		1	AT-22	(CF)
43-30550	43-30560	11	C-87	(CF)
43-30561		1	AT-22	(CF)
43-30562	43-30568	7	C-87	(CF)
43-30569	43-30571	3	C-87A	(CF)
43-30572	43-30573	2	C-87	(CF)
43-30574		1	AT-22	(CF)
43-30575	43-30583	9	C-87	(CF)
43-30584		1	AT-22	(CF)
43-30585	43-30627	43	C-87	(CF)
44-10253	44-10752	500	B-24J	(CF)
44-28061	44-28276	216	B-24J	(NT)
44-39198	44-39298	101	C-87	(CF)
44-40049	44-41389	1,341	B-24J	(CO)
44-41390	44-41806	417	B-24L	(CO)
44-41807	44-42722	916	B-24M	(CO)
44-44049	44-44501	453	B-24J	(CF)

Serial Range		Number	Type/Model	
44-48753		1	XB-24N	(FO)
44-48754	44-49001	248	B-24J	(FO)
44-49002	44-50251	1,250	B-24L	(FO)
44-50252	44-51928	1,677	B-24M	(FO)
44-52053	44-52059	7	YB-24N	(FO)
44-52978	44-52987	10	C-87	(CF)
		18,482		

Navy Cognisance Aircraft

Serial Range		Number	Type/Model	
59350	60009	660	PB4Y-2	(CO)
66245	66324	80	PB4Y-2	(CO)
90020	90050	31	RY-3	(CO)
90057	90059	3	RY-3	(CO)
		774		
		18,482		
		19,256		

Appendix F: B-24 Serial Numbers by Manufacturer and Block Number

All Liberators prior to serial 41-23640 were produced before adoption of the Block (or, in the case of CAC, 'Series') numbering system. Those aircraft are listed sequentially in Appendix E and are not repeated here.

Consolidated, San Diego (Co)

Model	Serial Range	Number
B-24D-1	41-23640—41-23668	29
	41-23671—41-23693	23
	41-23697—41-23724	28
	41-23728—41-23750	23
B-24D-5	41-23751—41-23755	5
	41-23759—41-23790	32
	41-23794—41-23824	31
B-24D-7	41-23825—41-23849	25
	41-23853—41-23858	6
B-24D-10	41-23864—41-23902	39
	41-23906—41-23919	14
B-24D-13	41-23920—41-23958	39
	41-23960—41-23969	10
B-24D-15	41-23970—41-24003	34
	41-24007—41-24026	20
	41-24030—41-24099	70
B-24D-20	41-24100—41-24138	39
	41-24142—41-24157	16
	41-24164—41-24171	8
	41-24175—41-24219	45
B-24D-25	41-24220—41-24311	92
	41-24339	1
B-24D-30	42-40058—42-40137	80
B-24D-35	42-40138—42-40217	80

Model	Serial Range	Number
B-24D-40	42-40218—42-40257	40
B-24D-45	42-40258—42-40322	65
B-24D-50	42-40323—42-40344	22
B-24D-53	42-40345—43-40392	48
B-24D-55	42-40393—42-40432	40
B-24D-60	42-40433—42-40482	50
B-24D-65	42-40483—42-40527	45
B-24D-70	42-40528—42-40567	40
B-24D-75	42-40568—42-40612	45
B-24D-80	42-40613—42-40652	40
B-24D-85	42-40653—42-40697	45
B-24D-90	42-40698—42-40742	45
B-24D-95	42-40743—42-40787	45
B-24D-100	42-40788—42-40822	35
B-24D-105	42-40823—42-40867	45
B-24D-110	42-40868—42-40917	50
B-24D-115	42-40918—42-40962	45
B-24D-120	42-40963—42-41002	40
B-24D-125	42-41003—42-41047	45
B-24D-130	42-41048—42-41092	45
B-24D-135	42-41093—42-41137	45
B-24D-140	42-41138—42-41172	35
B-24D-145	42-41173—42-41217	45
B-24D-150	42-41218—42-41257	40

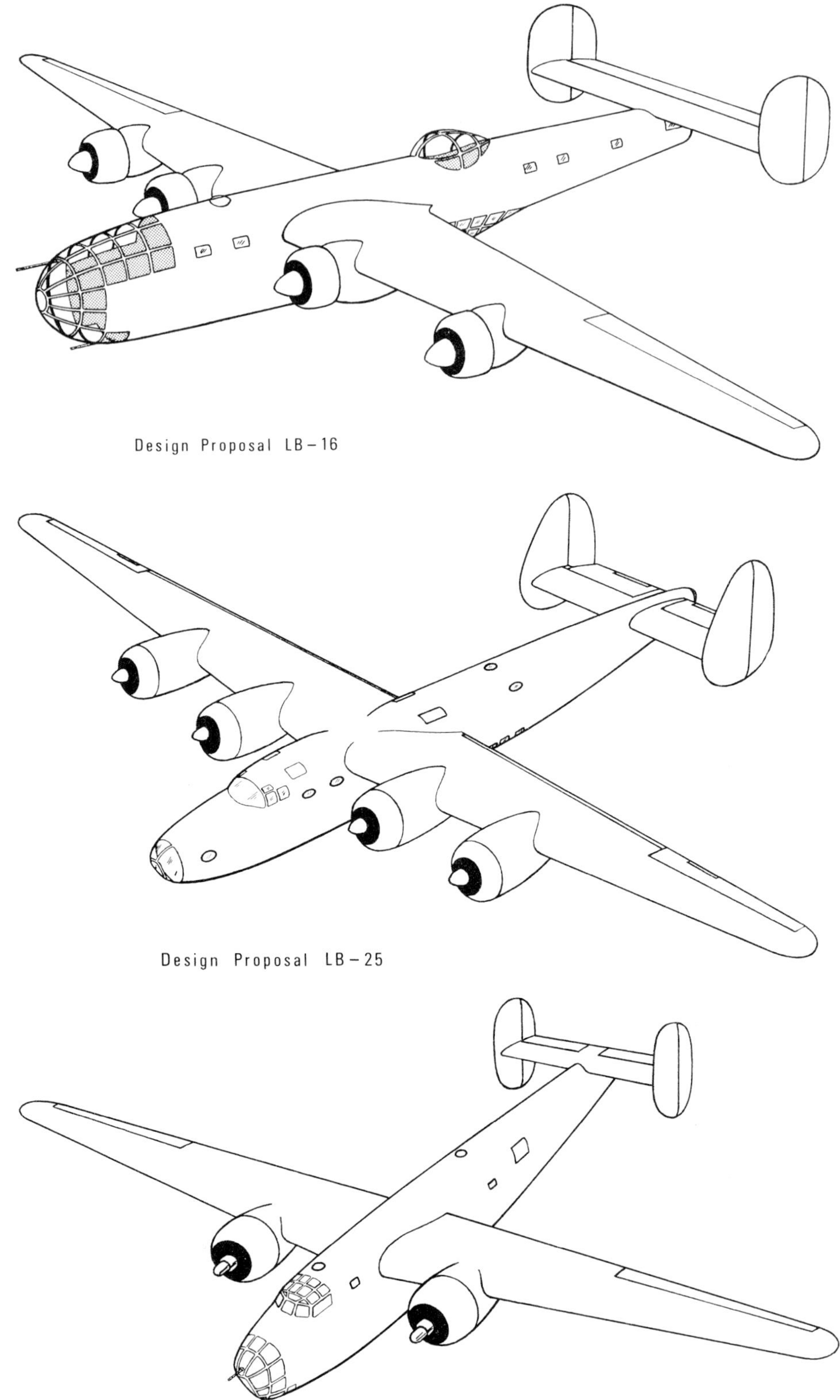

Design Proposal LB–16

Design Proposal LB–25

Design Proposal LB–29

Model	Serial Range	Number
B-24D-155	42-72765—42-72814	50
B-24D-160	42-72815—42-72864	50
B-24D-165	42-72865—42-72914	50
B-24D-170	42-72915—42-72963	49
B-24J-1	42-72964—42-73014	51
B-24J-5	42-73015—42-73064	50
B-24J-10	42-73065—42-73114	50
B-24J-15	42-73115—42-73164	50
B-24J-20	42-73165—42-73214	50
B-24J-25	42-73215—42-73264	50
B-24J-30	42-73265—42-73314	50
B-24J-35	42-73315—42-73364	50
B-24J-40	42-73365—42-73414	50
B-24J-45	42-73415—42-73464	50
B-24J-50	42-73465—42-73514	50
B-24J-55	42-99936—42-99985	50
B-24J-60	42-99986—42-100035	50
B-24J-65	42-100036—42-100085	50
B-24J-70	42-100086—42-100135	50
B-24J-75	42-100136—42-100185	50
B-24J-80	42-100186—42-100235	50
B-24J-85	42-100236—42-100285	50
B-24J-90	42-100286—42-100335	50
B-24J-95	42-100336—42-100385	50
B-24J-100	42-100386—42-100435	50
B-24J-105	42-109789—42-109838	50
B-24J-110	42-109839—42-109888	50
B-24J-115	42-109889—42-109938	50
B-24J-120	42-109939—42-109988	50
B-24J-125	42-109989—42-110038	50
B-24J-130	42-110039—42-110088	50
B-24J-135	42-110089—42-110138	50
B-24J-140	42-110139—42-110188	50
B-24J-145	44-40049—44-40148	100
B-24J-150	44-40149—44-40248	100
B-24J-155	44-10249—44-40348	100
B-24J-160	44-40349—44-40448	100
B-24J-165	44-40449—44-40548	100
B-24J-170	44-40549—44-40648	100
B-24J-175	44-40649—44-40748	100
B-24J-180	44-40749—44-40848	100
B-24J-185	44-40849—44-40948	100
B-24J-190	44-40949—44-41048	100
B-24J-195	44-41049—44-41148	100
B-24J-200	44-41149—44-41248	100
B-24J-205	44-41249—44-41348	100
B-24J-210	44-41349—44-41389	41
B-24L-1	44-41390—44-41448	59
B-24L-5	44-41449—44-41548	100
B-24L-10	44-41549—44-41648	100
B-24L-15	44-41649—44-41748	100
B-24L-20	44-41749—44-41806	58
B-24M-1	44-41807—44-41848	42
B-24M-5	44-41849—44-41948	100
B-24M-10	44-41949—44-42048	100
B-24M-15	44-42049—44-42148	100
B-24M-20	44-42149—44-42248	100
B-24M-25	44-42249—44-42348	100
B-24M-30	44-42349—44-42448	100
B-24M-35	44-42449—44-42548	100
B-24M-40	44-42549—44-42648	100
B-24M-45	44-42649—44-42722	74

Consolidated, Fort Worth (CF)

Model	Serial Range	Number
B-24D-1	42-63752—42-63796	45
B-24D-5	42-63797—42-63836	40
B-24D-10	42-63837—42-63896	60
B-24D-15	42-63897—42-63971	75
B-24D-20	42-63972—42-64046	75
B-24E-10	41-29009—41-29023	15
B-24E-15	41-29024—41-29042	19
B-24E-20	41-29043—41-29061	19
B-24E-25	41-29062—41-29115	54
	42-64395—42-64431	37
B-24H-1	41-29116—41-29187	72
	42-64432—42-64440	9
B-24H-5	41-29188—41-29258	71
	42-64441—42-64451	11
B-24H-10	41-29259—41-29335	77
	42-64452—42-64501	50
B-24H-15	41-29336—41-29606	271
B-24H-20	41-29607—41-29608	2
	42-50277—42-50354	78
B-24H-25	42-50355—42-50410	56
B-24H-30	42-50411—42-50451	41
B-24J-1	42-64047—42-64141	95
B-24J-5	42-64142—42-64236	95
B-24J-10	42-64237—42-64328	92
	42-64330—42-64346	17
B-24J-12	42-64329	1
B-24J-15	42-64347—42-64394	48
	42-99736—42-99805	70
B-24J-20	42-99806—42-99871	66
B-24J-25	42-99872—42-99935	64
B-24J-30	44-10253—44-10302	50
B-24J-35	44-10303—44-10352	50
B-24J-401	42-50452—42-50508	57
B-24J-40	44-10353—44-10374	22
B-24J-45	44-10375—44-10402	28
B-24J-50	44-10403—44-10452	50
B-24J-55	44-10453—44-10502	50
B-24J-60	44-10503—44-10552	50
B-24J-65	44-10553—44-10602	50
B-24J-70	44-10603—44-10652	50
B-24J-75	44-10653—44-10702	50
B-24J-80	44-10703—44-10752	50
B-24J-85	44-44049—44-44148	100
B-24J-90	44-44149—44-44248	100
B-24J-95	44-44249—44-44348	100
B-24J-100	44-44349—44-44448	100
B-24J-105	44-44449—44-44501	53

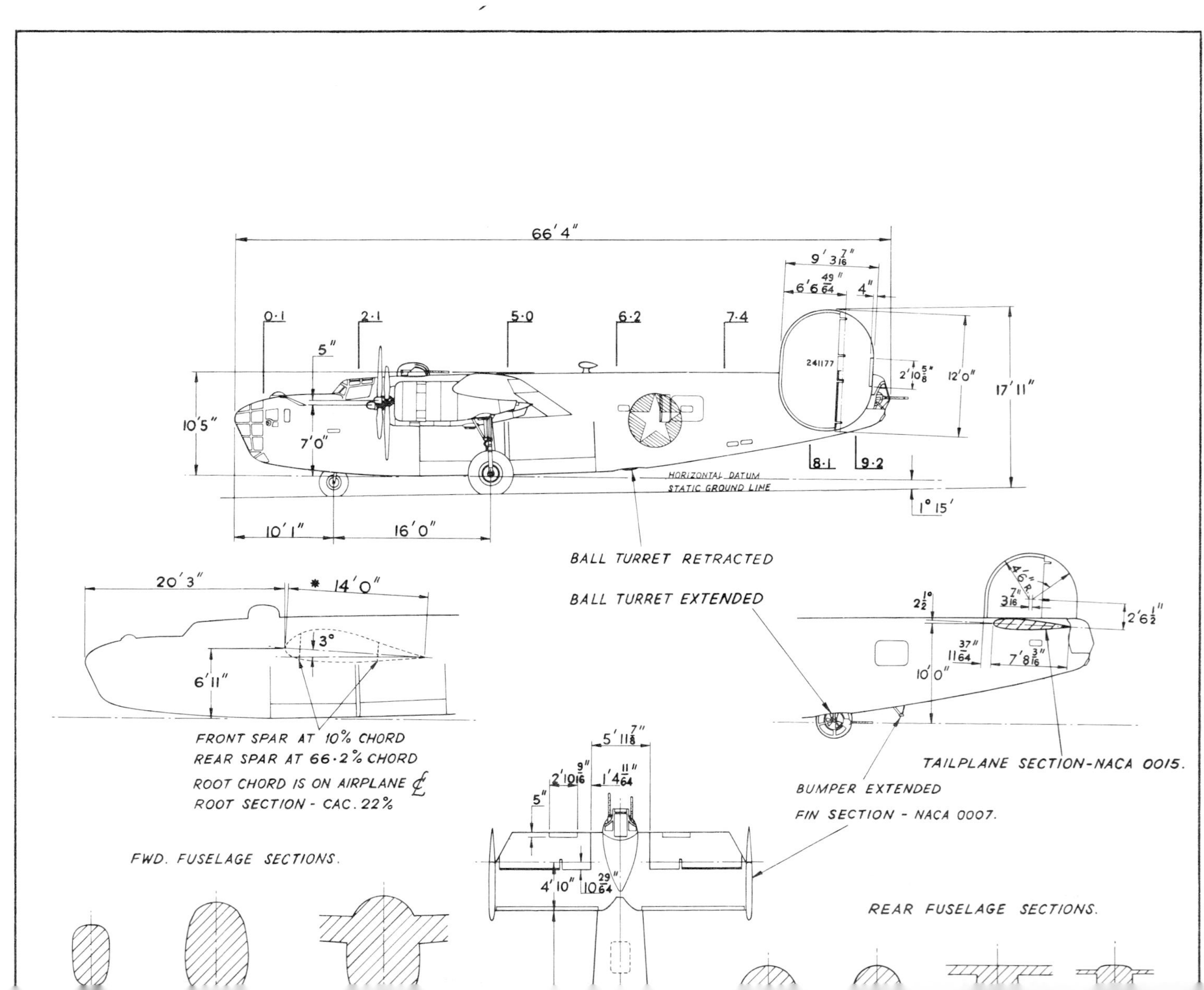
66′ 4″
9′ 3 7/16″
6′ 6 49/64″
4″
0·1
2·1
5·0
6·2
7·4
5″
241177
2′ 10 5/8″
12′ 0″
17′ 11″
10′5″
7′0″
8·1
9·2
HORIZONTAL DATUM
STATIC GROUND LINE
1° 15′
10′ 1″
16′ 0″
BALL TURRET RETRACTED
BALL TURRET EXTENDED
20′ 3″
* 14′ 0″
3°
6′ 11″
FRONT SPAR AT 10% CHORD
REAR SPAR AT 66·2% CHORD
ROOT CHORD IS ON AIRPLANE ℄
ROOT SECTION - CAC. 22%
FWD. FUSELAGE SECTIONS.
4′ 6″ R.
3 7/16″
2 1/2°
2′ 6 1/2″
11 37/64″
7′ 8 3/16″
10′ 0″
5′ 11 7/8″
2′ 10 9/16″
1′ 4 11/64″
5″
4′ 10″
10 29/64″
TAILPLANE SECTION-NACA 0015.
BUMPER EXTENDED
FIN SECTION - NACA 0007.
REAR FUSELAGE SECTIONS.

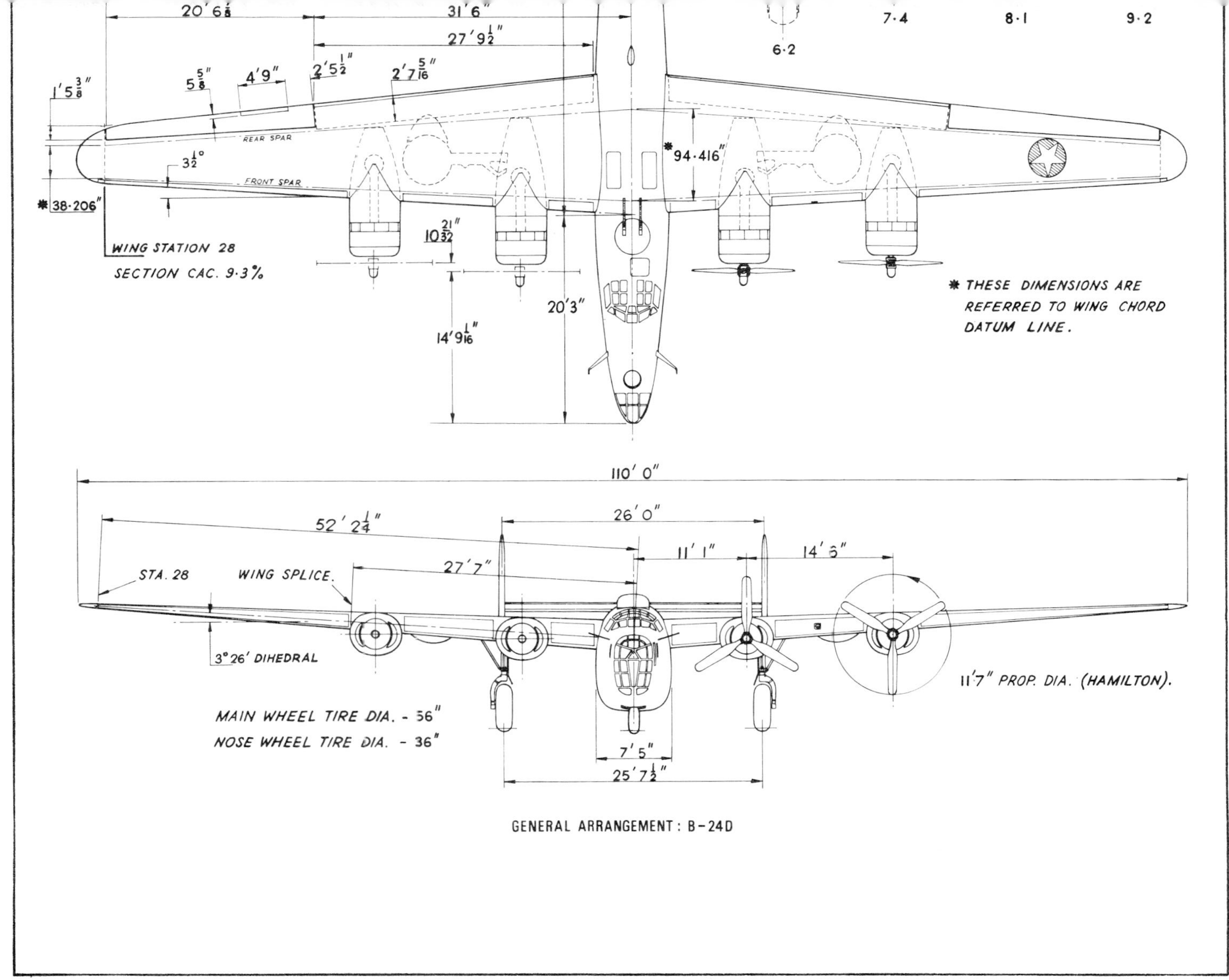

GENERAL ARRANGEMENT: B-24D

Ford, Willow Run (FO)

Model	*Serial Range*	*Number*
B-24E	42-7770	1
B-24E-1	42-6976—42-7005	30
B-24E-5	42-7006—42-7065	60
B-24E-10	42-7066—42-7122	57
B-24E-15	42-7123—42-7171	49
B-24E-20	42-7172—42-7229	58
B-24E-25	42-7230—42-7464	235
B-24H-1	42-7465—42-7717	253
B-24H-5	42-7718—42-7769	52
	42-52077—42-52113	37
B-24H-10	42-52114—42-52302	189
B-24H-15	42-52303—42-52776	474
	42-94729—42-94794	66
B-24H-20	42-94795—42-95022	228
B-24H-25	44-95023—42-95288	266
B-24H-30	42-95289—42-95503	215
B-24J-1	42-95504—42-95628	125
	42-50509—42-50759	251
B-24J-5	42-50760—42-51076	317
	42-51431—42-51610	180

Model	*Serial Range*	*Number*
B-24J-10	42-51611—42-51825	215
B-24J-15	42-51826—42-52075	250
B-24J-20	42-52076	1
	44-48754—44-49001	248
B-24L-1	44-49002—44-49251	250
B-24L-5	44-49252—44-49501	250
B-24L-10	44-49502—44-49751	250
B-24L-15	44-49752—44-50001	250
B-24L-20	44-50002—44-50251	250
B-24M-1	44-50252—44-50451	200
B-24M-5	44-50452—44-50651	200
B-24M-10	44-50652—44-50851	200
B-24M-15	44-50852—44-51051	200
B-24M-20	44-51052—44-51251	200
B-24M-25	44-51252—44-51451	200
B-24M-30	44-51452—44-51928	477
XB-24N	44-48753	1
YB-24N-1	44-52053—44-52059	7

Douglas, Tulsa (DT)

Model	Serial Range	Number
B-24D	41-11754—41-11756	3
B-24D	41-11864	1
B-24D-1	41-23725—41-23727	3
B-24D-5	41-23756—41-23758	3
B-24E	41-29007—41-29008	2
B-24E-1	41-28409—41-28416	8
B-24E-10	41-28417—41-28444	28
B-24E-15	41-28445—41-28476	32
B-24E-20	41-28477—41-28500	24
B-24E-25	41-28501—41-28573	73
B-24H-1	41-28574—41-28639	66

Model	Serial Range	Number
B-24J-1	42-51226—42-51292	67
B-24J-5	42-51293—42-51395	103
B-24J-10	42-51396—42-51430	35
B-24H-5	41-28640—41-28668	29
B-24H-10	41-28669—41-28752	84
B-24H-15	41-28753—41-28941	189
B-24H-20	41-28942—41-29006	65
	42-51077—42-51103	27
B-24H-25	42-51104—42-51181	73
B-24H-30	42-51182—42-51225	44

North American, Dallas (NT)

Model	Serial Range	Number
B-24G	42-78045—42-78069	25
B-24G-1	42-78070—42-78074	5
B-24G-5	42-78075—42-78154	80
B-24G-10	42-78155—42-78314	160
B-24G-15	42-78315—42-78352	38

Model	Serial Range	Number
B-24G-16	42-78353—42-78474	122
B-24J-1	42-78476—42-78794	319
B-24J-2	42-78475	1
B-24J-5	44-28061—44-28276	216

Appendix G: Liberator Constructor's Numbers (C/N's)

San Diego (CO)

Series	*Serial No.*	*C/N*
XB-24 (XB-24B)	39-680	1
LB-30A	AM258/263	1-6
LB-30B	AM910/929	1-20
YB-24	40-702	1
B-24A	40-2369/2377	1-9

Series	*Serial No.*	*C/N*
LB-30	AL503/AL641, FP685	1-140
B-24C	40-2378/2386	1-9
B-24D	40-696/701	1-6
B-24D	40-2349/2368	7-26
B-24D	41-1087/1142	27-82

Series	Serial No.	C/N	Series	Serial No.	C/N
B-24D	41-11587/11938	83-434	B-24J	42-99936/100435	3085-3584
B-24D	41-23640/24339	435-1134	B-24J	42-109789/110188	3585-3984
B-24D	42-40058/41257	1135-2334	B-24J, L, M	44-40049/42722	3985-6658
B-24D, J	42-72765/73514	2335-3084			

Fort Worth (CF)

Series	Serial No.	C/N	Series	Serial No.	C/N
B-24D, J	42-63752/64394	1-643	B-24H	41-29212/29258	261-307
B-24J	42-99736/99935	644-843	B-24H	42-64452/64501	308-357
B-24J	44-10253/10752	844-1343	B-24H	41-29259/29608	358-707
B-24J	44-44049/44501	1344-1796	B-24H, J	42-50277/50508	708-939
B-24E	41-29009/29115	1-107	C-87	42-107249/107275	1-27★
B-24E, H	42-64395/64440	108-153	C-87	43-30548/30627	28-107
B-24H	41-29116/29211	154-249	C-87	44-39198/39298	1-101
B-24H	42-64441/64451	250-260	C-87	44-52978/52987	102-111

★Originally CO 1107-1133

Willow Run (FO)

Series	Serial No.	C/N	Series	Serial No.	C/N
B-24E, H	42-6977/7769	1-793	B-24J	42-51431/52076	2962-3607
B-24H	42-52077/52776	794-1493	XB-24N	44-48753	3608
B-24H, J	42-94729/95628	1494-2393	B-24J, L, M	44-48754/51928	3609-6783
B-24J	42-50509/51076	2394-2961	YB-24N	44-52053/52059	6908-6914

Tulsa (DT)

Series	Serial No.	C/N	Series	Serial No.	C/N
B-24E, H	41-28409/29006	1-598	B-24H, J	42-51077/51430	599-952

Dallas (NT)

Series	Serial No.	C/N	Series	Serial No.	C/N
B-24G, J	42-78045/78794	1-750	B-24J	44-28061/28276	751-966

Appendix H: Serial Numbers of B-24's Converted to F-7, F-7A and F-7B

Serial Range		Number	Series	Serial Range		Number	Series
41-11653		1	XF-7	42-64165		1	F-7A
42-40433		1	F-7	42-64167	42-64174	8	F-7A
42-40476		1	F-7	42-64175		1	F-7A★
42-40488		1	F-7	42-64176	42-64181	6	F-7A
42-40494		1	F-7	42-64184	42-64186	3	F-7A
42-64047	42-64056	10	F-7A	42-64188	42-64194	7	F-7A
42-64102	42-64106	5	F-7A	42-64197		1	F-7A
42-64158		1	F-7A				

★This aircraft was later converted to a CB-24J.

Serial Range		*Number*	*Series*	*Serial*		*Number*	*Series*
42-64199	42-64201	3	F-7A	44-40663		1	F-7B
42-64203		1	F-7A	44-40669		1	F-7B
42-64235	42-64238	4	F-7A	44-40847		1	F-7B
42-64239		1	F-7B	44-40883		1	F-7B
42-64240	42-64242	3	F-7A	44-40885		1	F-7B
42-64243		1	F-7B	44-40895		1	F-7B
42-64245		1	F-7A	44-40961		1	F-7B
42-64246		1	F-7B	44-40963		1	F-7B
42-64247	42-64248	2	F-7A	44-40967		1	F-7B
42-64249	42-64250	2	F-7B	44-41013		1	F-7B
42-64251		1	F-7A	44-41040		1	F-7B †
42-64253	42-64254	2	F-7A	44-41477		1	F-7B
42-64255	42-64256	2	F-7B	44-41678	44-41681	4	F-7B
42-64260		1	F-7B	44-41943	44-41944	2	F-7B
42-64262		1	F-7B	44-41968		1	F-7B
42-64323		1	F-7A	44-42026		1	F-7B
42-64331		1	F-7A	44-42028		1	F-7B
42-64337		1	F-7A	44-42031		1	F-7B
42-64369		1	F-7A	44-42059		1	F-7B
42-64371		1	F-7A	44-42097	44-42098	2	F-7B
42-73020		1	F-7A	44-42135	44-42136	2	F-7B
42-73028		1	F-7A	44-42141		1	F-7B
42-73031		1	F-7A	44-42239		1	F-7B
42-73033	42-73035	3	F-7A	44-42265		1	F-7B
42-73038	42-73045	8	F-7A	44-42272		1	F-7B
42-73047	42-73050	4	F-7A	44-42302		1	F-7B
42-73052	42-73053	2	F-7A	44-42313	44-42314	2	F-7B
42-73132		1	F-7A	44-42328		1	F-7B
42-73157		1	F-7A	44-42350		1	F-7B
44-40147		1	F-7B	44-42373		1	F-7B
44-40160		1	F-7B	44-42376		1	F-7B
44-40197	44-40199	3	F-7B	44-42387		1	F-7B
44-40209		1	F-7B	44-42403		1	F-7B
44-40356		1	F-7B	44-42409		1	F-7B
44-40376		1	F-7B	44-42426		1	F-7B
44-40412	44-40413	2	F-7B	44-42452		1	F-7B
44-40415	44-40417	3	F-7B	44-42478	44-42480	3	F-7B
44-40419		1	F-7B	44-42580		1	F-7B
44-40422	44-40423	2	F-7B	44-42604	44-42605	2	F-7B
44-40602		1	F-7B	44-42621		1	F-7B
44-40612		1	F-7B	44-42653		1	F-7B
44-40616		1	F-7B	44-42686	44-42687	2	F-7B
44-40622		1	F-7B	44-42689	44-42691	3	F-7B
44-40625	44-40627	3	F-7B	44-42693		1	F-7B
44-40629		1	F-7B	44-42695	44-42722	28	F-7B
44-40656		1	F-7B	44-51518		1	F-7B
44-40658	44-40659	2	F-7B	44-51655		1	F-7B

† 44-41040 may have been converted to F-7B configuration at one time but conclusive evidence is lacking..

Appendix I: Serial Numbers of B-24's Converted to C-109

42- 7721
42-51368
390
411
420
424-427
429
615
647
659
676
684
697
706
712
716
721
727
730
734
740
748
756
758
766
774
782
784
786
788
792-793
809-810
817
825-826
830
839
844
846-847
42-51849-850
854
857
860
862
876-877
883
887
890
893
901
904
921
930
962
982-983
42-52000-001
005-006
012
014
020-021
023
033
042
049
44-48755
792
877
879
882-883
888
890
892
948
968
974
979
984
44-48995-996
999
44-49001
007-009
011-020
022-023
025
030-031
034-035
037
040
045-046
050-051
057
059-060
062-063
065
067
071
075
077
079
184
191
197
204
208
219
222
230
234-236
238
240
245-249
251
253
255-258
265
44-49267
269-272
274-277
280-281
283-285
288-290
292
295
299
302-303
305
313
317
319
326
330
333
337
344
348
351-354
358-359
450
466
490
510
615
621
628
660
662
684
691
704
715
720
723
728

Appendix J: B-24/PB4Y-1 Serial No/Bureau No Equivalents

BuNo	*AAF Serial*	*BuNo*	*AAF Serial*
31936-937	41-23826-827	31946-948	41-24049-051
938-939	926-927	949	088
940-941	946-947	950	053
942-945	993-996	951-952	083-084

BuNo	AFF Serial	BuNo	AAF Serial
31953	41-24087	32056-060	42-40571-575
954	086	061-063	578-580
955-958	115-118	064	568
959-962	131-134	065	576
963-965	176-178	066	581
966-968	208-210	067	584
969-971	240-242	068	720
972-973	271-272	069	709
974-978	303-307	070	714
979-983	42-40083-087	071	712
984-988	118-122	072	668
989-993	151-155	073	711
994-998	187-191	074	715
999	228	075	717
		076	721
32000-003	224-227	077	724
004-008	259-263	078	718
009	258	079	723
010	264	080-084	725-729
011-014	270-273	085	785
015-018	275-278	086	860
019	274	087	794
020	257	088	802
021	279	089-090	797-798
022-023	282-283	091	796
024-028	295-299	092	799
029	390	093-094	805-806
030	423	095	847
031	281	096	868
032	429	097-098	880-881
033	432	099	889
034-036	434-436	100-101	883-884
037	439	102	882
038	430	103	888
039-040	437-438	104	890
041	440	105-106	892-893
042	442	107	895
043	204	108	898
044	425	109	902
045	441	110	899
046	443	111	903
047	446	112	906
048-049	561-562	113-114	908-909
050	565	115	923
051	567	116	932
052	566	117	936
053	395	118	943
054	570	119-120	947-948
055	564	121-122	951-952

BuNo	*AAF Serial*
32123	42-40950
124	953
125	42-41001
126	42-40971
127	42-41016
128-129	019-020
130-131	024-025
132-133	037-038
134	045
135	035
136	034
137	036
138	039
139	042
140	044
141	046
142	048
143	055
144	052
145	102
146	119
147	121
148	123
149-150	130-131
151	138
152	122
153	134
154	139
155	148
156	153
157-159	165-167
160	204
161	42-72883
162-163	888-889
164	893
165-166	922-923
167-168	925-926
169	42-73036
170-189	091-110
190-202	168-180
203-212	205-214
213-222	400-409
223-227	42-100026-030
228-232	161-165
233-241	296-304
242-246	378-382
247-250	417-420
251-260	42-109905-914
261-270	955-964

BuNo	*AAF Serial*
32271-280	42-110007-016
281-290	124-133
291-300	174-183
301-310	44-40054-063
311-320	074-083
321-324	094-097
325-334	174-183
335	214
38733-741	44-40215-223
742-751	254-263
752-761	304-313
762-771	344-353
772-781	384-393
782	42-78271
783-791	44-40445-453
792-801	504-513
802-811	574-583
812-821	634-643
822-831	694-703
832-841	764-773
842-851	834-843
852-861	904-913
862-866	968-972
867-871	44-41034-038
872-876	100-104
877-881	166-170
882-891	239-248
892-901	279-288
902-913	297-308
914-923	319-328
924-933	359-368
934-944	406-416
945-954	484-493
955-964	519-528
965-979	554-568
46725-727	44-41601-603
728-736	683-691
737	701
63915	42-40483
916	762
917	809
918	843
919	(unknown)
920	42-63794★
921	42-40867
922	42-40901

BuNo	AAF Serial	BuNo	AAF Serial
63923	42-40824	63951	42-72919
924	789	952	918
925	907	953	928
926	761	954	924
927	792	955	921
928	897	956	920
929	816	957	914
930	628	958	917
931	861	959	916
932	810		
933	553	65287-391	44-41702-806
934	846	392-396	857-861
935	42-63795★		
936	42-40811	90132	44-41862
937	757	133	44-42017
938	876	134-135	44-41864-865
939	825	136-191	887-942
940	42-63796★	192-194	997-999
941	42-63799★	195-211	44-42000-016
942	42-72905	212-231	032-051
943	909	232-258	067-093
944	915	259-271	176-188
945	897		
946	903	90462-466	44-41569-573
947	906	467-483	584-600
948	895		
949	913		
950	910		

★ These aircraft known to be PB4Y-1's but sequence of BuNo's shown may be incorrect.

Appendix K-1: AAF Serial Numbers of Liberators allocated to the RAF under Lend Lease

B-24's

40-2350	41-11610-612	41-11681-685	42-40098-099
41-1087	619	687	124
1093	621	690	136
1096-097	626-628	692	138
1107-112	633-635	694	148
1114-115	643-645	696-697	150
1119-122	647-649	699-703	157-163
1124	651	705	284-289
1126-127	654	710-716	291-294
1129	658-662	718	300-310
41-11589-590	664-668	720-727	444-445
594	670	734-738	448-449
599	672	740-741	454-455
604-607	679	748-756	457-459

42-40462-465
467
546-548
556
558-559
569
582-583
585-587
589-593
599
625
633
641
643
652
672
674
678
693
696
698
700
703
42-63807-865
867-892
898
901-902
911-952
992-999
42-64000-036
057-060
062-063
065-100
121 126
128-152
161
187
204-233
244
252
259
261
263-322
324-330
332-336
338
345
352

42-64355
357
360
366
375-379
381-384
386-387
432
42-99736-745
747
750-751
753
756-757
765
767
769
774-794
796-797
800
804
806
809-812
814-815
817-848
850
852
857
859
861
863-935
42-100349
383-384
386-392
395
397
399
44-10253
256-436
438-439
441-483
644-752
44-44049-119
130-349
360-500
44-48798
859
885
44-49026

44-49049
052
056
058
061
064
066
069
072
076
078
081
083
090
092-111
113-122
126
133
137
139-140
142
144-145
147-148
150-151
153-156
158-160
162-171
173
176
178
195-196
199-200
203
205-207
209-210
212
214-215
217-218
220-221
223
426
439-441
443-444
447
449
452-453
455
457

44-49459
462
465
470
473
475
477
671
682
686
689
692
697
701
706
709
713
717
722
729
755-758
760-762
765-766
768-770
774-775
777-778
780-782
785
787
790
792-794
798-800
802-806
812-813
815
817
819-821
823-825
828
830-836
842-843
846
856
861-862
871
874
883
888

B-24's contd.

44-49891	44-49992-994	44-50086-088	44-50171-172
893	996-998	099-102	174-179
898	44-50001	104	181
901	003	107	183-185
903	008	112	187
906	010	114	191-193
909	013-014	116	195-199
917	016-017	118	202
922	019-021	120-121	206
931	023	123-125	208
934	025-027	127-128	210
951	029	130-131	212-213
956	032-039	134-135	215
959	041-047	137-140	217
962	049	147-148	226-228
968-969	051-053	151	230
978-979	055-057	153-155	232
982	059-061	157-161	234
985	063-079	163-165	242
987	081-082	167-168	246

Note: Officially, 355 B-24L's were supplied under Lend Lease. The two extra B-24L serials shown above were allocated as replacements for aircraft that were destroyed before delivery to the UK.

C-87's
44-39219-226
236-237
248-261

Appendix K-2: B-24 Theatre Transfers to the RAF

European Theatre of Operations. (*ETO*)

41-28868	42-52771
41-29568	42-94771
42-50744	42-94797
42-51226	42-94813
42-51350	42-94847
42-51529	42-94856
42-52483	42-94981
42-52572	44-10421★
42-52573	44-10533
42-52591	44-10574
42-52620	44-10594
42-52681	44-10597
42-52712	44-10611
42-52731	44-40380
42-52766	44-40457

Mediterranean Theatre of Operations (*MTO*)

41-11906	42-78110
41-28722†	42-78113†
41-29278†	42-78129
42-52205†	42-78143†
42-64341†	42-78144†
42-78080†	42-78153†
42-78096†	

★Returned to US control, April 1945.
†Returned to US control after VE Day.

Appendix L: Liberators in RAAF Service

RAAF A-72	*USAAF*	*RAAF* A-72	*USAAF*	*RAAF* A-72	*USAAF*	*RAAF* A-72	*USAAF*
-1	41-11904	-75	44-41403	-132	44-41507	-189	44-41982
-2	41-23720	-76	44-41402	-133	44-41509	-190	44-41983
-3	41-24018	-77	44-41404	-134	44-41510	-191	44-41984
-4	42-40489	-78	44-41405	-135	44-41511	-192	44-41987
-5	42-40512	-79	44-41444	-136	44-41512	-193	44-41988
-6	42-40522	-80	44-41450	-137	44-41513	-194	44-41989
-7	41-24070	-81	44-41454	-138	44-41514	-195	44-41990
-8	41-24290	-82	44-41452	-139	44-41658	-196	44-41991
-9	41-11868	-83	44-41455	-140	44-41665	-197	44-41992
-10	41-24127	-84	44-41456	-141	44-41677	-198	44-41993
-11	42-40514	-85	44-41457	-142	44-41682	-300	44-28063
-12	42-41132	-86	44-41458	-143	44-41828	-301	44-28064
-13	42-100194	-87	44-41459	-144	44-41831	-302	44-28065
-31	44-40251	-88	44-41460	-145	44-41834	-303	44-28066
-32	44-40252	-89	44-41575	-146	44-41836	-304	44-28067
-33	44-40410	-90	44-41579	-147	44-41885	-305	44-28069
-34	44-40406	-91	44-41580	-148	44-41886	-306	44-28087
-35	44-40408	-92	44-41581	-149	44-41515	-307	44-28097
-36	44-40409	-93	44-41582	-150	44-41516	-308	44-28099
-37	44-40407	-94	44-41583	-151	44-41517	-309	44-28101
-38	44-40411	-95	44-41604	-152	44-41518	-310	44-28104
-39	44-40653	-96	44-41653	-153	44-41530	-311	44-28139
-40	44-40649	-97	44-41608	-154	44-41531	-312	44-28086
-41	44-40652	-98	44-41639	-155	44-41532	-313	44-28125
-42	44-40651	-99	44-41640	-156	44-41533	-314	44-28127
-43	44-40654	-100	44-41609	-157	44-41529	-315	44-28256
-44	44-40650	-101	44-41654	-158	44-41971	-316	44-28079
-45	44-40870	-102	44-41611	-159	44-41972	-317	44-28122
-46	44-40871	-103	44-41577	-160	44-41949	-318	44-28070
-47	44-40872	-104	44-41612	-161	44-41950	-319	44-28082
-48	44-40873	-105	44-41617	-162	44-41951	-320	44-28132
-49	44-41194	-106	44-41605	-163	44-41952	-321	44-28089
-50	44-41197	-107	44-41607	-164	44-41953	-322	44-28076
-51	44-41203	-108	44-41618	-165	44-41954	-323	44-28103
-52	44-41207	-109	44-41630	-166	44-41955	-324	44-28138
-53	44-41204	-110	44-41656	-167	44-41957	-325	44-28107
-54	44-41196	-111	44-41663	-168	44-41958	-326	44-28118
-55	44-41206	-112	44-41664	-169	44-41959	-327	44-28123
-56	44-41193	-113	44-41629	-170	44-41960	-328	44-28092
-57	44-41205	-114	44-41633	171	44-41961	-329	44-28100
-58	44-41195	-115	44-41632	-172	44-41962	-330	44-28114
-59	44-41389	-116	44-41631	-173	44-41963	-331	44-28068
-60	44-41388	-117	44-41634	-174	44-41964	-332	44-28131
-61	44-41375	-118	44-41635	-175	44-41965	-333	44-28098
-62	44-41249	-119	44-41636	-176	44-41956	-334	44-28091
-63	44-41374	-120	44-41642	-177	44-41966	-335	44-28081
-64	44-41384	-121	44-41666	-178	44-41967	-336	44-28108
-65	44-41385	-122	44-41637	-179	44-41969	-337	44-28078
-66	44-41373	-123	44-41638	-180	44-41970	-338	44-28095
-67	44-41386	-124	44-41657	-181	44-41973	-339	44-28124
-68	44-41376	-125	44-41505	-182	44-41974	-340	44-28096
-69	44-41391	-126	44-41499	-183	44-41975	-341	44-28113
-70	44-41392	-127	44-41502	-184	44-41976	-342	44-28137
-71	44-41393	-128	44-41503	-185	44-41977	-343	44-28080
-72	44-41394	-129	44-41504	-186	44-41978	-344	44-28106
-73	44-41401	-130	44-41506	-187	44-41979	-345	44-28084
-74	44-41395	-131	44-41508	-188	44-41980	-346	44-28126

RAAF A-72	USAAF	RAAF A-72	USAAF	RAAF A-72	USAAF	RAAF A-72	USAAF
-347	44-28109	-362	44-28073	-377	44-28174	-392	44-28178
-348	44-28072	-363	44-28149	-378	44-28171	-393	44-28075
-349	44-28077	-364	44-28168	-379	44-28143	-394	44-28159
-350	44-28090	-365	44-28156	-380	44-28145	-395	44-28180
-351	44-28074	-366	42-78787	-381	44-28151	-396	44-28172
-352	44-28135	-367	44-28088	-382	44-28183	-397	44-28175
-353	44-28133	-368	44-28094	-383	44-28142	-398	44-28158
-354	44-28083	-369	44-28136	-384	44-28162	-399	44-28150
-355	44-28085	-370	44-28144	-385	44-28179	-400	44-28147
-356	44-28130	-371	44-28164	-386	44-28155	-401	44-28153
-357	44-28129	-372	44-28140	-387	44-28062	-402	44-28160
-358	44-28093	-373	44-28148	-388	44-28169	-403	44-28161
-359	44-28111	-374	44-28071	-389	44-28167	-404	42-78726
-360	44-28128	-375	44-28102	-390	44-28166	-405	42-78727
-361	44-28134	-376	44-28121	-391	44-28146		

Not included in the above are two aircraft, 44-28105 and one other, which were lost in transit. At least forty-eight additional B-24J deliveries were cancelled in 1945.

A-72-1 through A-72-12 were theatre transfers. USAAF Lend Lease records also list 42-40508 and 42-40534 as theatre transfers to the RAAF but these do not appear on the RAAF serial list.

Appendix M-1: AAF Serial Numbers of B-24's Allocated to the RCAF under Lend Lease

41-24001	42-40470	44-10584★	44-10640	44-44351	44-50006
41-24236	42-40471	44-10585	44-10641	44-44352	44-50012
41-24277	42-40557	44-10586	44-10642	44-44353	44-50022
41-24281	42-40560	44-10587	44-10643	44-44354	44-50040
42-40231	42-64101	44-10588	44-44120	44-44355	44-50050
42-40447	42-64182	44-10589	44-44121	44-44356	44-50062
42-40450	42-64183	44-10590	44-44122	44-44357	44-50182
42-40451	42-99795	44-10592	44-44123	44-44358	44-50186
42-40452	44-10254	44-10593	44-44124	44-44359	44-50214
42-40453	44-10255	44-10634	44-44125	44-49112	44-50218
42-40456	44-10437	44-10635	44-44126	44-49127	44-50591
42-40460	44-10440	44-10636	44-44127	44-49131	44-50601
42-40461	44-10540	44-10637	44-44128	44-49157	44-50622
42-40466	44-10581	44-10638	44-44129	44-49161	44-50639
42-40469	44-10583	44-10639	44-44350	44-49198	

★44-10584 allocated but crashed in US; not delivered.

Appendix M-2: AAF Serial Numbers of RAF Liberators held in Canada under British Commonwealth Air Training Plan

42-99793	42-99817	42-99827	42-99898	42-99904	44-10391
42-99794	42-99820	42-99894	42-99899	44-10273	44-10392
42-99809	42-99821	42-99895	42-99900	44-10284	44-10396
42-99812	42-99822	42-99896	42-99902	44-10285	44-10397
42-99815	42-99823	42-99897	42-99903	44-10385	44-10399

44-10425	44-10435	44-10672	44-10737	44-44156	44-44309
44-10427	44-10436	44-10673	44-10738	44-44157	44-44310
44-10428	44-10483	44-10674	44-10739	44-44158	44-44311
44-10429	44-10670	44-10675	44-10740	44-44159	44-44312
44-10434	44-10671	44-10736	44-10741	44-44308	44-44313

Appendix N: Summary of Liberator Aircraft Modified at Modification Centres, 1942-5

		1942	*1943*	*1944*	*1945*	*Total*
American Airlines						
New York	AAF	25	—	—	—	25
	UK	74	—	—	—	74
		99	—	—	—	99
Bechtel-McCone-Parsons-						
Birmingham	AAF	—	1,061	2,551	188	3,800
	H2X	—	—	27	—	27
	UK	—	—	30	—	30
		—	1,061	2,608	188	3,857
Curtiss Wright						
Buffalo	AAF	—	1	—	—	1
Curtiss Wright						
Louisville	AAF	—	—	—	1	1
Consolidated						
Fort Worth	AAF	167	417	—	—	584
	USN	28	30	—	—	58
	UK	13	199	—	—	212
		208	646	—	—	854
Consolidated						
Louisville	AAF	—	15	90	78	183
	UK	—	117	922	470	1,509
	RCAF	—	3	35	16	54
		—	135	1,047	564	1,746
Consolidated						
Tucson	AAF	2	1,749	1,832	83	3,666
	USN	—	143	—	—	143
	UK	—	20	—	—	20
	RAAF	—	—	22	—	22
	F-7	—	—	1	36	37
		2	1,912	1,855	119	3,888

Douglas						
Tulsa	AAF	—	—	171	292	463
	H2X	—	—	66	—	66
	RAAF	—	—	—	113	113
		—	—	237	405	642
Lockheed						
Dallas	RAAF	—	17	—	—	17
	F-7	—	5	—	—	5
		—	22	—	—	22
Martin						
Omaha	F-7	—	—	18	—	18
Northwest Airlines						
St Paul	AAF	309	762	351	898	2,320
	H2X	—	—	661	—	661
	USN	—	10	—	—	10
	RAAF	—	—	—	1	1
	F-7	—	11	107	—	118
	F-7 H2X	—	—	—	4	4
		309	783	1,119	903	3,114
United Airlines						
Cheyenne	AAF	—	—	—	138	138
	Totals	618	4,560	6,884	2,318	14,380

Appendix O: Summary of Liberator/Privateer Performance

	Weights			Top speed at military power at combat weight	Service ceiling	Range in statute miles	
	Basic	*Combat*	*War max.*			*At max. cruise power with payload as shown*	*Ferry (no load)*
XB-24	27,500	38,361	46,100	273@15,000	31,500	2,850 (2,500 lb)	4,700
B-24A	30,000	41,000	56,000	293@15,000	30,500	2,200 (4,800 lb)	4,000
B-24C	32,050	41,000	56,000	313@25,000	34,000	2,100 (5,000 lb)	3,650
B-24D	37,000	56,000	71,200	303@25,000	32,000	1,800 (5,000 lb)	3,500
B-24E	34,000	41,000	67,700	302@25,000	30,000	2,500 (5,000 lb)	4,100
B-24G	38,000	56,000	71,200	290@25,000	28,000	1,700 (5,000 lb)	3,300
B-24H	38,000	56,000	71,200	290@25,000	28,000	1,700 (5,000 lb)	3,300
B-24J	38,000	56,000	71,200	290@25,000	28,000	1,700 (5,000 lb)	3,300
B-24L	37,500	56,000	71,200	290@25,000	28,000	1,700 (5,000 lb)	3,300
B-24M	37,500	56,000	71,200	290@25,000	28,000	1,700 (5,000 lb)	3,300
XB-24N	38,300	56,000	65,000	292@25,000	28,000	2,000 (5,000 lb)	3,500
PB4Y-1	(Comparable to AAF Series Equivalents)						
PB4Y-2	37,765	62,000	70,231	245@13,750	21,200	2,630 (4,000 lb)	3,800
RY-3	30,429	56,000	62,000	264@14,700	21,300	3,700 (6,200 lb)	n/a

Appendix P: USAAF Inventory of First Line B-24 Aircraft ★

	Total AAF	*CON US*	*ETO*	*MTO*	*POA*	*FEAF*	*CBI*	*Alaska*	*XX AF*	*Latin AM*	*Enroute & Other*
12/41	89	89	—	—	—	—	—	—	—	—	—
3/42	147	124	—	—	—	—	—	—	—	—	23
6/42	309	250	—	17	—	—	5	9	—	—	28
9/42	601	459	34	52	32	—	2	18	—	7	—
12/42	834	515	41	81	40	60	22	21	—	18	36
3/43	1,171	715	74	87	40	92	74	25	—	21	43
6/43	1,867	1,165	51	192	42	170	94	25	—	18	110
9/43	2,821	1,848	96	204	55	281	110	19	—	26	182
12/43	3,490	1,920	308	268	105	361	167	16	—	25	320
3/44	4,954	2,370	772	868	154	385	134	13	—	31	227
6/44	5,877	2,272	1,458	982	240	474	112	21	3	41	274
9/44	6,043	2,257	1,471	1,190	208	522	126	18	3	60	188
12/44	5,678	2,572	1,183	951	222	403	115	17	7	69	139
3/45	5,267	1,989	1,045	1,136	262	455	135	21	—	68	156
6/45	4,986	3,494	159	91	236	526	173	15	—	70	222
9/45†	3,458	2,489	19		704		48	12	—	72	114
12/45†	1,103	906	5		93		5	8	—	37	49

†ETO/MTO and POA/FEAF totals combined beginning 9/45.

★ Excludes all other Liberator variants.

Note: Column Abbreviations are as follows: CON US—Continental United States; ETO—European Theatre of Operations; MTO—Mediterranean Theatre of Operations; POA—Pacific Ocean Area; FEAF—Far East Air Forces; CBI—China/Burma/India; XXAF—Twentieth Air Force; Latin AM—Latin America

Appendix Q: US Navy Inventory of Liberator/Privateer Aircraft

		JAN	*FEB*	*MAR*	*APR*	*MAY*	*JUN*	*JUL*	*AUG*	*SEP*	*OCT*	*NOV*	*DEC*
1942	PB4Y-1	—	—	—	—	—	—	—	—	7	19	33	51
1943	PB4Y-1	63	58	81	77	135	143	177	222	221	266	282	277
	RY-1	—	—	—	—	—	—	—	—	—	—	—	1
1944	PB4Y-1	276	296	314	342	348	365	397	n/a	506	588	623	631
	RY-1	3	3	3	3	3	3	3	3	3	3	3	3
	RY-2	—	—	5	5	5	5	5	5	5	4	4	4
	PB4Y-2	—	—	2	2	7	20	36	n/a	37	57	96	140
1945	PB4Y-1	642	643	636	623	604	582	368	345	337	178	178	166
	RY-1	3	3	3	3	3	3	3	3	3	3	3	3
	RY-2	4	4	4	4	4	4	—	—	—	—	—	—
	PB4Y-2	189	242	288	312	394	411	448	499	550	651	651	653
	RY-3	1	1	1	1	1	1	1	1	1	10	10	17

★ Does not include aircraft accepted but not yet delivered—a figure which in 1944-5 often exceeded 150 per month. Also does not include the three XPB4Y-2's, one of which was dropped from inventory in June 1945 and a second in August 1945. The third was still on the lists as of December 1945.

Appendix R: The Liberator Power Plant

All Liberator aircraft were powered by some version of the Pratt & Whitney R-1830 Twin Wasp engine. The original design for this power plant was laid down in mid-1930, with the first 'hot run' being made on 16 April 1931 and the first delivery taking place on 11 August 1932. The Twin Wasp was a 14-cylinder, two-row, air-cooled radial of 1,830 in.^{3} (30 litres) displacement. The engine was approximately 63 in. in length and had a diameter of approximately 48·5 in. (different models varied slightly), giving a frontal area of about 12·8 ft^2. Bore and stroke were equal at 5·5 in. and compression ratio was 6·7:1. The three-piece crankcase was of forged aluminium alloy; the cylinders utilised steel barrels and cast aluminium alloy heads, and had one inlet and one exhaust valve per cylinder. Valves were pushrod-operated. The crankshaft was a one piece, two-throw, counterbalanced unit running in three roller-bearing supports. The Twin Wasp was a rugged and reliable engine that powered many other World War II aircraft, including the C-47, PBY and F4F. Because of the large demand, the automobile firms of Buick and Chevrolet were called in to supplement Pratt & Whitney production. This they did in large numbers. Total R-1830 production was 173,618 engines.

The following is a list of R-1830 'dash numbers' that were associated with the Liberator/Privateer family. All horsepower figures given are take-off ratings.

- -33 1,200 hp, 2:1 reduction gear ratio, 1,480 lb dry weight. Used in XB-24, YB-24 and B-24A.
- -41 1,200 hp, 2:1 ratio, 1,490 lb. High speed clutch removed for turbo. Used in XB-24B, B-24C.
- -43 1,200 hp, 16:9 ratio, 1,500 lb. Used in XB-24B, B-24D, B-24E, XB-24F, B-24G, B-24H, B-24J (few), PB4Y-1, C-87, XB-41. Scintilla magnetos.
- -43A Same as -43 except Bosch magnetos. Built by Buick and Chevrolet only. Used in B-24D, B-24E, C-87, B-24L and B-24M.
- -59 1,200 hp, 16:9 ratio, 1,575 lb. Only four built for test, in B-24D, of engine and supercharger cooling changes. Engines later converted to -94's.
- -61 1,200 hp, 16:9 ratio, 1,495 lb. This was commercial S3C4-G engine installed in Liberator aircraft under original British contract. When US took over LIB II's as LB-30's, engine required a military designation and S3C4-G became R-1830-61.
- -65 1,200 hp, 16:9 ratio, 1,500 lb. Same as -43 except different carbs. Built by Buick only. Used on B-24D, B-24J, B-24M, PB4Y-1.
- -65A Same as -43A except for different carbs. Built by Buick and Chevrolet only. Used in B-24J, B-24L, B-24M and C-87. Also installed in XB-24B.
- -69 Estimated 1,250 hp, 16:9 ratio, 1,520 lb. Planned as production version of -59 but superseded by -75. None built
- -75 1,350 hp, 16:9 ratio, 1,555 lb. Suitable for 130 octane gasoline. Plain main bearings. Built by Buick only. Used in XB-24N and YB-24N. Also installed in XB-24B and XB-24K.
- -94 1,350 hp, 16:9 ratio, 1,573 lb. Suitable for 130 octane. Used in PB4Y-2, RY-3 and Convair Model 39 (XR2Y-1).
- -98 To be same as -94 except plain main bearings. Planned for use in PB4Y-2 but none built.

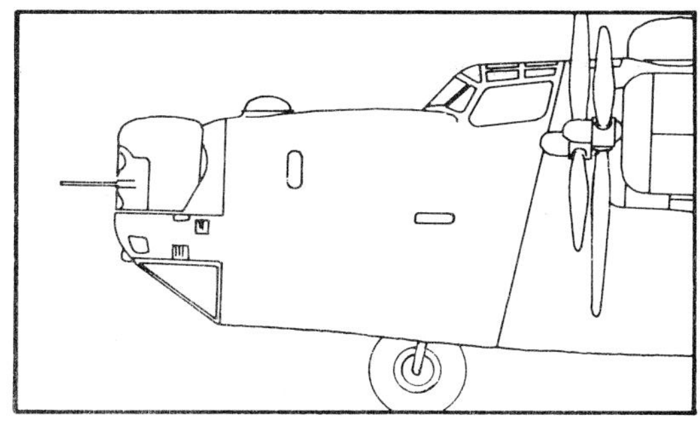

Bibliography

Allen, Hugh: *Goodyear Aircraft.* Cleveland, 1947.

Blue, Allan G.: 'LB-30's in Latin America', *Journal of the American Aviation Historical Society,* 1st Quarter, 1970.

Civilian Production Administration, Bureau of Demobilization, Special Study No. 21: *Aircraft Production Policies Under the National Defense Advisory Commission and Office of Production Management—May 1940 to December 1941.* Undated.

Craven, W. F., and J. L. Cate (eds): *The Army Air Forces in World War II.* 7 vols. Chicago, 1948-58.

Famme, J. H.: 'Design Analysis of Consolidated B-24 Liberator', *Aviation,* July 1945.

Feeny, William D.: *In Their Honor.* New York, 1963.

'Ferrets and How They Work', *Radar,* 20 August 1944.

Garfield, Brian: *The Thousand-Mile War.* New York, 1969.

Griffin, J. A.: *Canadian Military Aircraft.* Ottawa, 1969.

Haugland, Vern: *The AAF Against Japan.* New York, 1948.

Hay, T. Benson: 'This Wing May Win the War', *Saturday Evening Post,* 12 April 1941.

Hinton, H. B.: *Air Victory: The Men and the Machines.* New York, 1948.

Holley, I. B.: *Buying Aircraft: Materiel Procurement for the Army Air Forces.* Washington, 1964.

Keen, Harold: 'Mystery Airfoil', *Popular Aviation,* June 1940.

La Farge, Oliver: *The Eagle in the Egg.* Boston, 1949.

'Liberators Over Biscay', *Naval Aviation News,* February 1951.

Maurer, Maurer (ed.): *Air Force Combat Units of World War II.* Washington, 1961.

Millikan, Clark B.: *Report on Wind Tunnel Tests on a Davis Tapered Monoplane Wing and a Similar Consolidated Corporation Wing* (Galcit Report No. 201-B). Pasadena, 1937.

'The Navy's "Bat"', *Naval Aviation News,* June 1950.

'New Ferrets for the Jap Radar Search', *Radar,* 30 June 1945.

Robertson, Bruce: *British Military Aircraft Serials 1912-1963.* London, 1964.

Sherrod, Robert: *History of Marine Corps Aviation in World War II.* Washington, 1952.

Thiesmeyer, L. R., and J. E. Burchard. *Combat Scientists.* Boston, 1947.

Thompson, G. R. and others: *The Signal Corps: The Test (December 1941 to July 1943).* Washington, 1957.

USAAF: *Fifteenth Air Force Statistical Summary, 1943-1945.* 1945.

USAAF: *Statistical Summary of Eighth Air Force Operations, 17 Aug 1942-8 May 1945.*

USAAF Air Material Command: *B-24 Construction and Production Analysis—Ford, Willow Run.* Dayton, 1946.

USAAF Air Material Command: *B-24 Production and Construction Analysis—Consolidated—Vultee Aircraft Corporation, San Diego, California.* Dayton, 1946.

USAAF Office of Statistical Control: *Army Air Forces Statistical Digest—World War II.* 1945. Also Supplement No. 1 thereto. 1946.

US Court of Claims: *Davis Airfoils, Inc. v. United States.* 110 F. Supp. 460.

US Department of Commerce, Civil Aeronautics Administration: *U.S. Military Aircraft Acceptances 1940-1945.* Washington, 1946.

Wagner, Ray: *American Combat Planes.* New York, 1960.

'War Against the U-boat', *Radar,* September 1944.

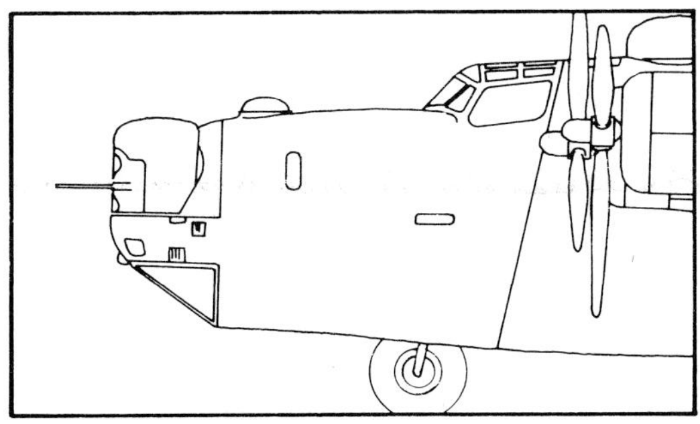

Index

Names in italics are those of individual aircraft; numbers in italics refer to illustrations; (N) indicates a footnote reference.